The Latin American Studies Book Series

The Latin American Studies Book Series promotes quality scientific research focusing on Latin American countries. The series accepts disciplinary and interdisciplinary titles related to geographical, environmental, cultural, economic, political, urban and health-related research dedicated to Latin America. The series publishes comprehensive monographs, edited volumes and textbooks refereed by a region or country expert specialized in Latin American studies.

The series aims to raise the profile of Latin American studies, showcasing important works developed focusing on the region. It is aimed at researchers, students, and everyone interested in Latin American topics.

Submit a proposal: Proposals for the series will be considered by the Series Advisory Board. A book proposal form can be obtained from the Publisher, Andrea Brody-Barre (andrea.brody-barre@springer.com).

Andrea Coronato · Grace B. Alves
Editors

Latin American Geomorphology

From the Crust to Mars

Springer

Editors
Andrea Coronato
Laboratorio de Geomorfología y Cuaternario
CADIC-CONICET
Ushuaia, Argentina

Grace B. Alves
Instituto de Geociências
Universidade Federal da Bahia
Salvador, Bahia, Brazil

ISSN 2366-3421 ISSN 2366-343X (electronic)
The Latin American Studies Book Series
ISBN 978-3-031-55177-2 ISBN 978-3-031-55178-9 (eBook)
https://doi.org/10.1007/978-3-031-55178-9

This Springer imprint is published by the registered company Springer Nature Switzerland AG
The registered company address is: Gewerbestrasse 11, 6330 Cham, Switzerland

Paper in this product is recyclable.

Foreword

One of the most important objectives of the International Association of Geomorphologists (IAG) is to support early-career geomorphologists from all around the World, especially those from economically disadvantaged countries. Since 2020, the International Association of Geomorphologists (IAG) has been celebrating International Geomorphology Week in the first week of March every year throughout the World. To overcome the problem of the time gap between the western and eastern extremes of the World and for wide participation, IAG has divided the whole World into different regions to organize the webinars. Central and South America, one of the regions, organized this type of event from the beginning. I am glad to see that from the presentations of the young geomorphologists in the IAG Central and South American Webinar during the Geomorphology Week, 2021, Andrea Coronato and Grace B. Alves have taken the initiative to edit a full volume. The International Association of Geomorphologists (IAG) appreciates their initiative. The IAG will be glad to see the formation of one or more Young Geomorphologists Groups in Central and South America.

The book contains seven chapters on varied geomorphological issues of different regions in Latin America. All of the papers have been contributed by young researchers covering areas from Costa Rica to southern Argentina, from mountains to seas and from the Caribbean to the Scotia lithological plates as well as planetary studies. The Chapter 2, written by María Romina Onorato, deals with the landforms and processes originated by modern tectonics along a lithological plate transform boundary in Tierra del Fuego, Argentina. The evolution of rock glaciers through glacial and periglacial processes in the upper Cochiguás catchment, Coquimbo region, Chile, has been analyzed in Chap. 3 by Catalina Pino Vargas. The megafan responses to paleoenvironmental changes during the late Quaternary in the Pantanal wetland, Brazil, have beautifully been explained by Fabiano do Nascimento Pupim et al. in Chap. 4. In Chap. 5, Breylla Campos Carvalho et al. elaborated on the impact of storm events on oceanic beaches between Ilha Grande and Cabo Frio, Rio da Janeiro, Brazil. There are some indirect impacts of landforms on cyclonic storms. It has beautifully been examined by Adolfo Quesada-Román in Chap. 6 on Costa Rica through geomorphic, hydrological, dendrochronological, and risk assessments to reduce the loss of lives and properties. New insights on Mars's water features in relation to illumination conditions in an exciting contribution by Mauro Spagnuolo et al. in Chap. 7. The volume covers the role of tectonic, glacial, and fluvial processes in geomorphic changes and the role of geomorphology on the cyclonic hazard. The last paper crossed the boundary of the Earth's surface to the extra-terrestrial sphere.

The book will likely attract the attention of the vast scientific community, including geomorphologists, environmentalists, environmental activists, and planners.

March 2023

Sunil Kumar De
President, International Association
of Geomorphologists (IAG)
Department of Geography
North-Eastern Hill University
Shillong, Meghalaya, India

Acknowledgements

The editors acknowledge Jorge O. Rabassa, Editor of the Springer Earth System Sciences, who kindly invited us to prepare this book after the first edition of the 1st Central and South America IAG webinar on March 2021.

Sunil Kumar De, President of the International Association of Geomorphologists (IAG), period 2022–2026, and former Vice-President (period 2017–2022), when the webinar took place, gently prepared the foreword.

We are greatly indebted to the specialists who kindly review the chapters published in this volume: Julieta Nóbile (CICTERRA-CONICET, Argentina), Patricio Becerra (Microcameras & Space Exploration S. A., Swiss), Jorge Rabassa (CADIC-CONICET, Argentina), Vanda Sales (Universidade Estadual do Vale do Acaraú-UVA, Brazil), and Ivan Bergier (EMBRAPA, Brazil).

We specially thank the corresponding authors whose contributions allow the preparation of this book.

About This Book

This book compiles the topics presented by young Latin American geomorphologists in the webinar held in March 2021 for Central and South America to celebrate the International Geomorphology Week promoted by the International Association of Geomorphologists (IAG).

The webinar took place over 6 hours including six presentations from young geomorphologists coming from Costa Rica, Brazil, Uruguay, and Argentina. They presented the research done for their Ph.D., post-doctoral, or young researchers' proposals related to geomorphology and related sciences as climatology, geology, or planetary studies.

The webinar had a wide audience from different countries of the region, even from Europe, who enriched the discussions with their contributions and questions.

The activity success encouraged the organization of later webinars in 2022, 2023 and 2024 as a way to maintain alive the interest of the young generations on the geomorphology of this vast and varied continent as well as to share new methods, results, and problems.

Contents

About the Editors

Andrea Coronato Dr. in Geography (University of Buenos Aires, Argentina, 1995). Full Researcher at CADIC-CONICET (Centro Austral de Investigaciones Científicas-Consejo Nacional de Investigaciones Científicas y Técnicas) in Ushuaia, Tierra del Fuego, Argentina. Professor at ICPA-UNTDF (Instituto de Ciencias Polares y Ambiente-Universidad Nacional de Tierra del Fuego, Antártida e Islas del Atlántico Sur), in Ushuaia, Argentina. IAG (International Association of Geomorphologists) Vice-President and Latin American co-opter (period 2022–2026), also former Latin American co-opter during period 2018–2022. Member of the Executive Committee of AACyG (Asociación Argentina de Cuaternario y Geomorfología), period 2022–2025 and former secretary during period 2012–2015. Specialist in Physical Geography, Geomorphology applied to Quaternary Geology and paleoenvironments of southern South America. Author of several chapters on glacial, periglacial, aeolian, and fluvial geomorphology of cold-semi arid environments of Patagonia and Tierra del Fuego, published in Springer Earth System Sciences and Springer Brief in Earth System Sciences volumes, among others. Author and co-author of more than 120 papers and book chapters, and co-author of two books published in international journals without interruptions since 1988. Professor of postgraduate field courses and congress field-trips in Tierra del Fuego and Patagonia organized by CADIC-CONICET and supported by IAG, NASA, among other institutions and academic societies. Promotor of science dissemination activities to primary, high-school, university, and postgraduate students, as well as to the general public, including the publication of articles in popular science journals and social media.

Grace B. Alves Dr. in Physical Geography (University of São Paulo, Brazil, 2014). Assistant Professor at the Geography Department of the UFBA (Federal University of Bahia), in Salvador, Bahia, Brazil, teaching and conducting research on landscape evolution, pedology, geomorphology, and geography. UGB (Geomorphological Union of Brazil) General Secretary and Coordinator of Colapso-Natureza e Sociedade Research Group. Specialist in pedology and geomorphology, focuses on the relationship between soil, landforms, and vegetation.

During her Ph.D., she delved into exploring the role of soil in landscape evolution, including its contribution to lake formation and the expansion of drainage networks in tropical conditions. She has published several articles in English and Portuguese on topics such as the geodiversity of Brazil, the physical properties of soils, and the landscape evolution. Additionally, she is a coordinator of the Genesis and Morphology of Soil Commission, belonging to the Núcleo Regional Nordeste of the SBCS (Brazilian Soil Science Society).

Chapter 1
Introduction

Andrea Coronato and Grace B. Alves

Abstract The Central and South American regions present diverse environments, landforms, and geomorphic processes that result from different processes produced in diverse climates, including equatorial, tropical, subtropical, temperate, and cold climates. These varied conditions have resulted in complex interactions among tectonic, climatic, and biotic factors, resulting in a wide array of landscapes. In this book, young geomorphologists from Central and South American countries present their research findings, showcasing the exciting topics and methods currently used in the field, presented during the 1st first webinar organized by the International Association of Geomorphologists in 2021. This research highlights the impact of endogenic forces, climatic events, and anthropic influences on the region's geomorphic processes and the potential for planetary geomorphology and remote sensing techniques to contribute to the field. This book emphasizes the importance of studying geomorphology to understand the region's history, evolution, and current dynamics to inspire scientists and contribute to active geomorphology in Central and South America.

Keyword Central America · South America · Young geomorphologists · Geomorphology

Central and South American regions encompass diverse environments, including tropical rainforests, high-altitude mountains, arid deserts, and coastal plains. For millions of years, these landscapes have been shaped by a complex interplay of

A. Coronato (✉)
Centro Austral de Investigaciones Científicas (CADIC-CONICET), Laboratorio de Geomorfología y Cuaternario, B. Houssay 200, 9410 Ushuaia, Argentina
e-mail: andrea.coronato@conicet.gov.ar

G. B. Alves
Universidade Federal da Bahia, Instituto de Geociências, Departamento de Geografia, R. Barão de Jeremoabo, s/n, 40170-290 Salvador, Bahia, Brazil
e-mail: alves.grace@ufba.br

A. Coronato and G. B. Alves (eds.), *Latin American Geomorphology*, The Latin American Studies Book Series, https://doi.org/10.1007/978-3-031-55178-9_1

tectonic, climatic, and biotic factors, resulting in various landforms and geomorphic processes. Studying these features, known as geomorphology, is critical for understanding the history, evolution, and current dynamics of these regions. In this book, we present the results of geomorphic research by young geomorphologists from Central and South American countries and show exciting topics and methods currently used. The chapters in the book result from presentations during the 1st webinar organized by the International Association of Geomorphologists (IAG) to celebrate International Geomorphology Week in Central and South America in March 2021.

Examples of different geomorphic processes and landforms that sculptured Central and South American spots are included, some generated by endogenic forces or triggered by climatic events or regimes. In addition, global advances in planetary geomorphology studies have encouraged young professionals to research combining physics and geomorphology through remote sensing techniques to replace fieldwork.

Central and South America is a vast portion of emerging lands of the western hemisphere, extending from 20°N to 55°S and covering approximately 20,425,000 km^2 (World Bank 2023a). They are supported by the Caribbean and South American lithospheric plates moving westward to the Cocos and Nazca plates, which move underneath by subduction. The Scotia and Antarctica plates developed transformation and convergence boundaries in the south.

Central American continental and marine reliefs result from the complex dynamics between the North and South American plates, in the middle of which the Caribbean originated. Inland tablelands, lacustrine basins, coastal plains, and volcanic arcs formed the main reliefs. Tectonic dynamics reflect processes triggered by trans-compressive-transcurrent borders on the north and south sides of the plate and by a collisional subduction border in the west (Cocos plate) and east. Active volcanism at these borders generates the most conspicuous relief in the continental and insulated territories.

South America has a complex origin. The territory began to set up during Precambrian times, when the Guyana, Brazil, and Patagonia shields formed in the Gondwana continent and moved westward during the Jurassic Meso-Atlantic range, separating the present American and African continents. Puna, Tandilia, and Ventania are old structures. The westward migration of the South American plate generates a collision with the border between the Nazca and Antarctic plates, causing uplifting and folding and forming the widest and highest section of the Andean Cordillera in Perú, Chile, and Argentina. During the Paleogene, the main orogenic episode lifted the Central America Mountains, the Andean Cordillera, and the Pampean Range in South America. Volcanism triggered by the subduction of the Pacific oceanic plates below South America generated volcanoes of various types almost along the entire Andes. Fissure volcanism formed tableland, cones, and maaric landscapes of varied ages eastward of the Andes. Between the oldest shields and the Cenozoic mountains, the extent and deep sedimentary basins of the Orinoco, Amazonas, and Paraguay-Paraná-Uruguay Rivers developed dense and active fluvial systems flowing from the Andes to the Atlantic. Low regions with impeded drainage in these units form extensive wetlands, such as the Pantanal, Bañados del Bermejo, and Ibera.

Undoubtedly, Andes Cordillera is the main geomorphological feature of the region. It spreads along 7000 km, mainly in the N-S direction, as a continental spine, where tectonism and volcanism are still active, increasing the elevation of the mountains. At the southern end, a west–east direction prevailed in the Fuegian Andes due to the formation of the Scotia plate between the South America and Antarctica plates. The average elevation is 3500 m a.s.l., although numerous summits over 6.000 m a.s.l. conform to the chain; the Aconcagua Mt (6960 m a.s.l.) and the Ojos del Salado, Pissis, and Bonete Chico Volcanoes, over 6769 m a. s.l., are the highest summits. High plains, such as the Bolivian Altiplano or Puna (3600 m a. s.l.), constitute one of the highest inhabited lands in the world. The lower summits (1200 m s.n.m.) are located in the Patagonian and Fuegian Andes.

The equatorial line extends along the northern portion of the region, configuring the domain of warm and humid climates in coastal and mountainous regions. The continuous low-pressure system attracts air from the northeast and southeast, forming the winds known as Alisios, which originated in the permanent High-Pressure Systems (HPS) located at 30° in latitude north and south in the Pacific and Atlantic oceans. The westerlies provide humid air masses that move according to relief. The insolation-temperature variations according to latitude, general atmospheric circulation, and variable distances to the sea are the main factors for South American climate diversity, from equatorial, tropical, subtropical, and temperate climates to cold temperate and cold climates.

After the World Bank (2023b), the population in Central and Latin America has only reached 32 inhabitants/m^2. Most people live in urban areas, mainly in cities with national or provincial governments and industries. The rural population is dispersed in small towns, villages, and farmhouses of various extents, some larger than 100 ha. The population is deeply related to agriculture, mining, and forestry activities or is settled in historical or cultural places.

Lithologies, tectonism, and climates result in the inexhaustible morpho-structural richness of Central and South America, which hosts magnificent components of biodiversity. Depending on the scale of analyses, multiple or single endogenic and exogenic processes combined with atmospheric dynamics offer a wide variety of landscapes where the population lives. This book humbly presents a spot from the perspective of young professionals (Fig. 1.1).

Quaternary global changes, such as atmospheric cooling and warming cycles, postglacial sea-level changes, solar activity variations, active tectonism, and volcanism, are extensively represented by glacial, periglacial, and aeolian systems, mass wasting complexes, upraised marine and lacustrine beaches, alluvial sedimentary plains, megafan development, upraised and sinking terrains where rapids and waterfalls are low, impeded drainage from wetlands or peatland areas, hard-rock tablelands, and exhumated terrains. Besides the present climatic change, the modification of landform and sediment distribution caused by anthropic activities, such as the extent of farming and cattle raising, mining, forestry, urban expansion over flood plains, terraces, mountain slopes, coasts, river channelization, and water reservoirs, among others, trigger some natural processes that affect populations more frequently and severely than in past decades. The present geomorphic process reflects the high

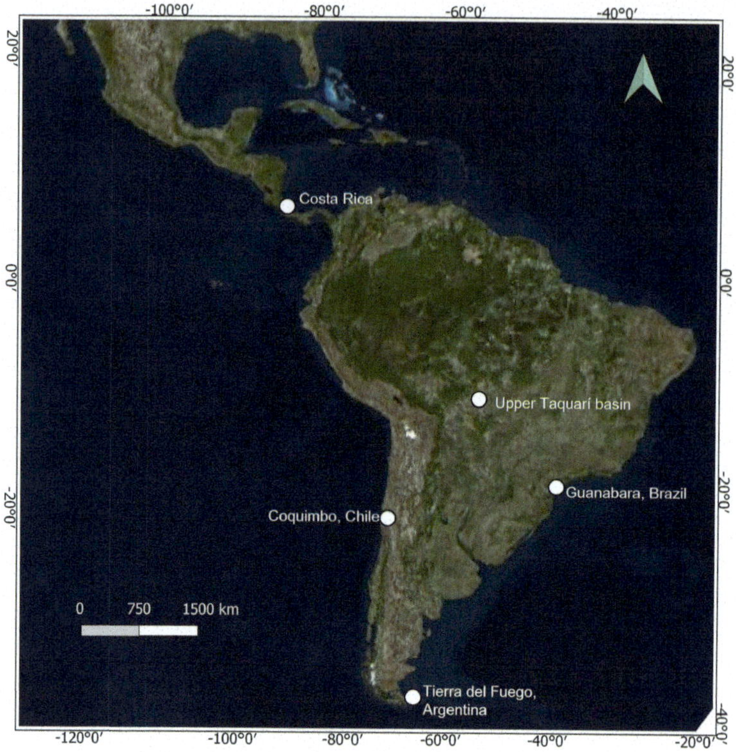

Fig. 1.1 Location of the studied areas in each chapter of this book. Image source Bing image mosaic from SAS. Planet platform, free availability (http://www.sasgis.org downloaded in March 2023)

dynamism of the earth and atmospheric system over a continent of multiple relief types, where the spatial spread of population and natural resource exploitation is expanding.

How does neotectonic activity along the transform boundary of the South America and Scotia plates continue to trigger seismic activity and form surficial fractures and scarps, collapses or sinking holes, or changes in river catchment design and geomorphic work? Will the rock glaciers persist as freshwater reservoirs in arid mountains in the context of climatic change in the Andean summit systems? Can anthropic activities in Pantanal wetlands be planned according to a dynamic fluvial system? Are numerical models and remote sensing techniques sufficiently efficient to replace field databases for comprehending coastal storms along the long Atlantic coastlines? Along the same line, could multidisciplinary procedures replace the lack of instrumental climatic and hydrological records to offer planning tools to prevent the effects of flooding and mass-wasting processes in the population when low-pressure atmospheric events occur? How much planetary geomorphology knowledge will advance according to progress in remote sensing technologies?

These are some of the questions derived from the chapters in this book. Undoubtedly, they will contribute to maintaining active geomorphology as Earth Science is deeply engaged in society. We hope that these concerns also awaken new challenges and motivate young scientists to discover the fascinating aspects of Central and South American geomorphology.

References

World Bank (2023a) Agricultural land (km^2)—Latin America & Caribbean. https://datos.bancomundial.org/indicador/ag.srf.totl.k2?end=2020&locations=ZJ&most_recent_value_desc=false&start=2020&view=bar

World Bank (2023b) Population density (people per sq. km of land area)—Latin America & Caribbean. https://datos.bancomundial.org/indicator/EN.POP.DNST?locations=ZJ

Chapter 2
Evidence of Neotectonism in Landforms and Landscape in the East of Tierra del Fuego, Argentina

María Romina Onorato

Abstract The Magallanes-Fagnano Fault System (MFFS) is the most important Quaternary structure in the Southern Andes. This transform system between the Scotia and South American plates extends over >600 km from the Atlantic Ocean to the Pacific Ocean and crosses Tierra del Fuego from W to E. This chapter is based on the analysis of satellite imagery and digital elevation models in combination with reconnaissance flights and field observations. Fieldwork includes validation of remote mapping along with reconnaissance, classification, and the identification of landforms and morphostructural features in selected areas. The geomorphological observations and the morphostructural analysis presented in this chapter argue that the MFFS generated ruptures and co-seismic displacements associated with the latest seismic events of 1949, among other episodes. This neotectonic evidence not only confirms the MFFS as the main potential seismogenic source for Tierra del Fuego but also shows the neotectonic influence on the transform boundary and demonstrates its impact on the landforms and landscape of Tierra del Fuego.

Keywords Morphotectonic · Neotectonic · Magallanes-Fagnano fault system

2.1 Introduction

This chapter presents investigation results on the influence of neotectonics in a plate boundary region affected by alpine and continental engulfment during the Quaternary. The study aims to analyze the tectonic-geomorphological evolution of glacial landforms, and the relationship between tectonic, erosional, and sedimentary processes.

The study area corresponds to the southern sector of South America, specifically to Isla Grande de Tierra del Fuego. It belongs to a transition area between the Central

M. R. Onorato (✉)
INGEO-UNSJ-CONICET, Gabinete de Neotectónica y Geomorfología, Av. Ignacio de la Roza 590 (O), Complejo Universitario "Islas Malvinas", J5402DCS Rivadavia, San Juan, Argentina
e-mail: onoratomariaromina@gmail.com; mronorato@unsj-cuim.edu.ar

A. Coronato and G. B. Alves (eds.), *Latin American Geomorphology*, The Latin American Studies Book Series, https://doi.org/10.1007/978-3-031-55178-9_2

Fuegian Andes Cordillera and its striations in the Fuegian fold and thrust belt domain (Borrello 1972; Olivero and Malumián 2008; Torres Carbonell et al. 2017).

Isla Grande de Tierra del Fuego is divided in an E-W direction by the transform boundary between the South American (to the north) and Scotia (to the south) plates, forming the MFFS (Winslow 1982). The latter runs from the western end of the Strait of Magellan to the limit of the Argentine Continental Shelf. It then continues eastward to the north of Isla de los Estados and along the North Scotia Ridge and reaches the northern end of the South Sandwich Islands archipelago (Forshyth 1975; Pelayo and Wiens 1989; Klepeis 1994) (Fig. 2.1a). The landforms analyzed are affected by the tectonic structures associated with the Magallanes-Fagnano Fault System (MFFS). More precisely, the studies have been conducted in the east sector of Tierra del Fuego (Fig. 2.1a, b).

The aim of this chapter is to expose the evidence of neotectonic activity in the landscape and the landforms. Landforms and landscape evolution are conditioned by both the predominant structural style in the study area and the tectonic activity during the Quaternary. Field morphological evidence for active faulting consists of relief forms resulting from recent tectonic activity of the MFFS, which denote the existence of plate boundary structures.

Active transcurrent faulting produces a characteristic assembles of landforms including scarps, linear valleys and rivers, sag ponds and pull-aparts, pressure and shutter ridges, progressive unconformities, wind gaps, tectonic gutters, unpaired terraces, offset drainage, river diversion, and radial drainage, among others (Vedder and Wallace 1970; Audemard 2003).

The main geomorphological expression structure of the MFFS along the plate boundary is recognized in the Chilean territory of the Island to the W from evidence recorded in the Quaternary alluvial cover that displays scarp and vegetation alignment, offset drainage, geometric shape in river patterns, and sag ponds. To the east, in the Argentine sector of the Island, the main structure of the MFFS is located near the northern shore of Fagnano Lake (Lodolo et al. 2003). Previous work by Onorato et al. (2019a, b, 2021) has analyzed morphotectonic evidence in a transect from lake Fagnano to the Atlantic coast and described fault scarps alternating their free face to the north and south, linear valleys and rivers, drainage control, offset drainage, wind gaps and drainage inversion, behead meanders, sag ponds and pull-aparts, and linear fault ridges (Figs. 2.1a and 2.2).

Recent studies describing the MFFS from a geometric and morphotectonic approach suggest that the fault system is characterized by different fault sections. Such fault sections have been defined based on the morphological features of scarps, natural outcrops, and the diverse evidence recorded (Roy et al. 2019; Sandoval and Pascale 2020; Onorato et al. 2021).

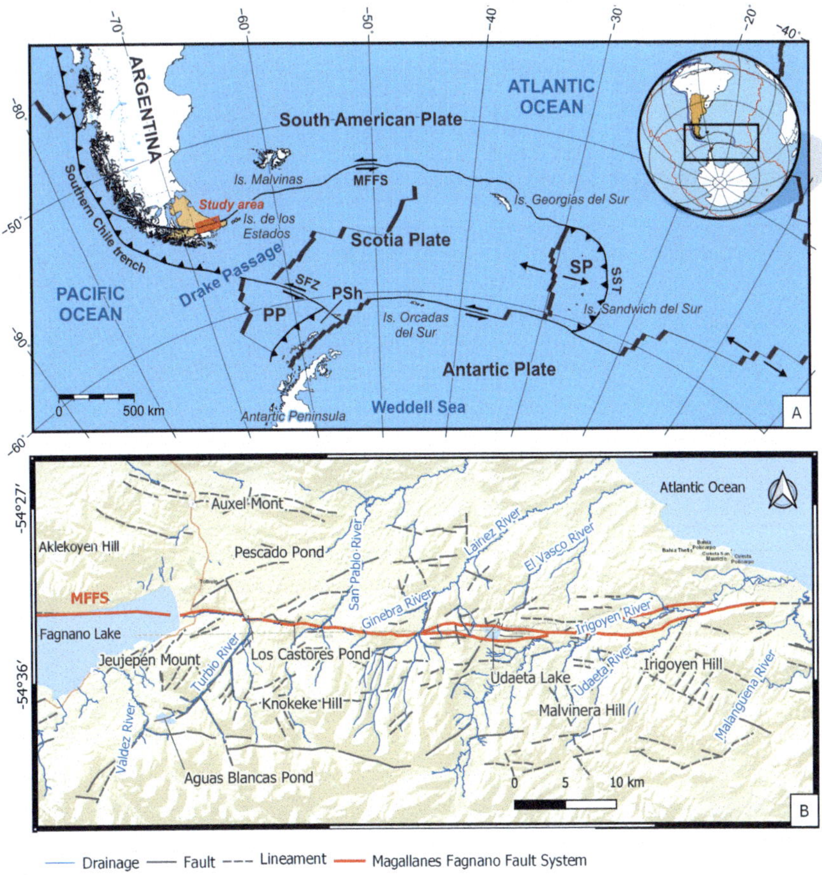

Fig. 2.1 **a** Tectonic setting of Tierra del Fuego (orange), based on data by Smalley et al. (2003). Abbreviations: MFFS: Magallanes-Fagnano Fault System; SFZ, Shackleton Fracture Zone; PSh: Shetlands Plate; PP: Phoenix Plate; SP: Sandwich Plate. **b** Location of the main trace of the eastern portion of Magallanes-Fagnano fault system is pointed by the red line, main rivers, and mountains of the area of interest

2.2 Climate and Vegetation, Quaternary Geology

Climate on Isla Grande de Tierra del Fuego is determined by its latitudinal position and insularity. Thus, it is cold humid oceanic climate in the center and south of Tierra del Fuego and cold sub-humid oceanic climate sub-humid oceanic cold toward the north (Coronato et al. 2017). In particular, the average temperature in the center of the island varies between 10 °C in summer (January) and 0 °C in winter (July).

The Fuegian Andes are the only mountain system extending west-eastward along the Andes. They form a topographic barrier to the circulation of southwest-westerly winds, resulting in a progressive rain shadow east of the Andes (Musotto et al. 2017).

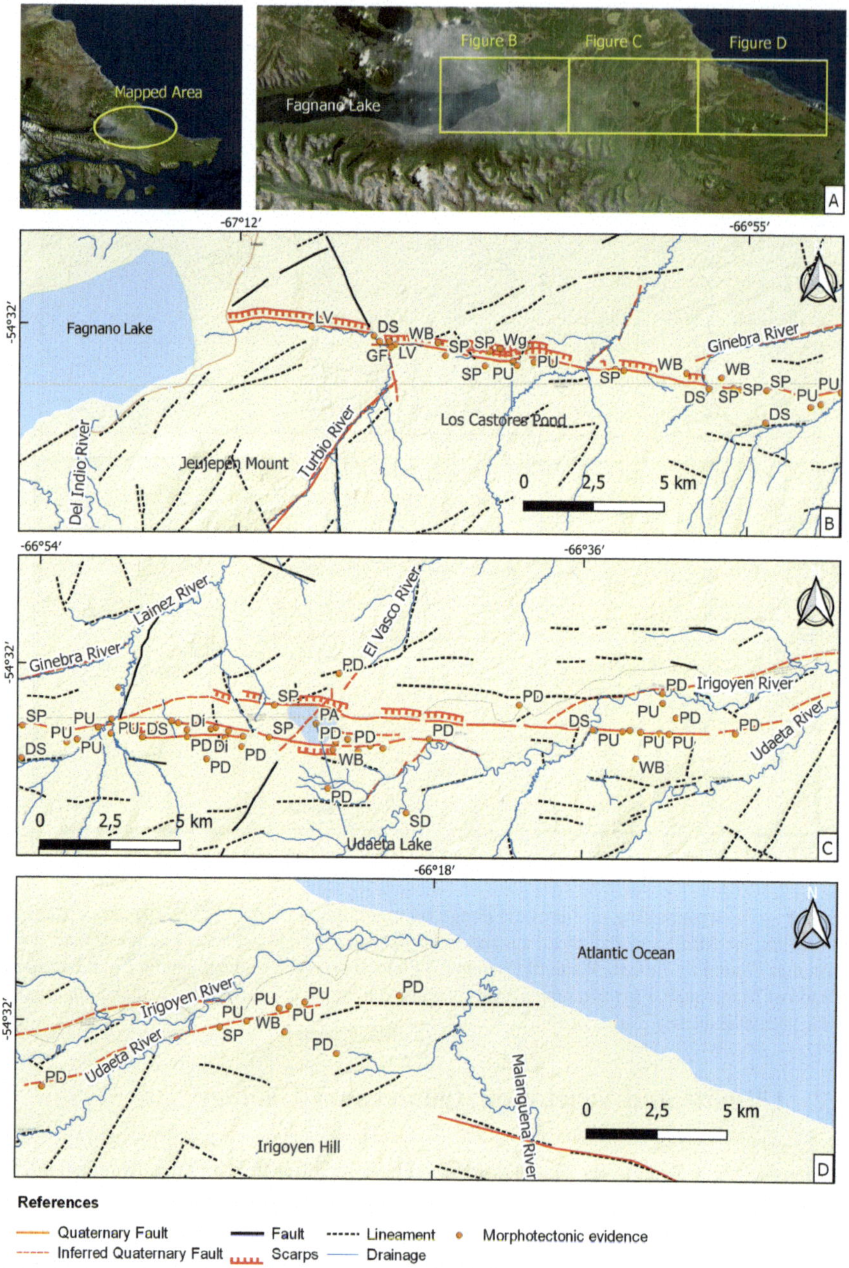

Fig. 2.2 Strip map of the morphotectonic evidence of the eastern portion of the main trace of Magallanes-Fagnano fault system, from longitude 67°12′W to longitude 66°18′W. The morphotectonic evidence mentioned in the text is marked with orange points. References: LV: Linear valley and rivers, Psk: Pseudokars, DS: Deflected streams, Wg: Wind gaps, SP: Sag Ponds, PA: Pull apart, WB: Whaleback, PU: Pop-up structure, DI: Drainage inversion, PD: Parallel patron drainage

The main annual precipitation ranges in central Tierra del Fuego vary from 400 to 500 mm and snowfall generally occurs between May and October.

The oceanic climate on the island is influenced by the strong winds from the west present throughout the year. The current vegetation shows a strong trans-Andean climatic and topographic gradient. To the north of Tierra del Fuego, the vegetation is characterized by the Fuegian-Patagonian Steppe, followed to the south successively by the Sub-Antarctic Deciduous Forest and the Sub-Antarctic Evergreen Forest. The grassland steppe, dominated by shrub and stunted vegetation, develops where the mean annual rainfall is <400 mm. In the south and west of Tierra del Fuego, the Sub-Antarctic Deciduous Forest is represented by two species: *Nothofagus pumilio* and *Nothofagus antarctica.* These are distributed up to an average altitude of 500–600 m a.s.l. and become dominant when precipitation exceeds 450 mm/year. On the southern and southeastern coast, precipitation increases above 700 mm/year and the Subantarctic Evergreen Forest develops, dominated by *Nothofagus betuloides* accompanied by *Drimys winteri*, Maytenus magellanica, and abundant ferns and mosses (Musotto et al. 2017).

In the Late Pleistocene, the Last Glacial Maximum (LGM) reached its maximum expansion about 25,000 years B.P. The retreat of the large glaciers during the Late Glacial began 15,000–16,000 14C years ago (between 20,000 and 18,000 calendar years) and lasted until the beginning of the Holocene, approximately 10,000 14C years ago (11,000 calendar years) (Rabassa et al. 2000). The study of glaciations in Patagonia and Tierra del Fuego has been featured in several publications since 1833 after Charles Darwin visited South America and the Pacific Ocean Islands aboard the Beagle.

Pleistocene glaciation extended extensively in the territory of Tierra del Fuego. Large glaciers from the Darwin Range covered Tierra del Fuego and flowed north, south, west, and east until they reached the present submarine shelf of the Atlantic Ocean (Meglioli et al. 1990; Isla and Schnack 1995). As a consequence, large and deep glacial valleys were generated in places that were partly flooded by the sea during the Holocene and are known today as the Strait of Magellan, the Bahía Inútil-Bahía San Sebastián depression, the Seno Almirantazgo-Lago Fagnano depression, the Carbajal Valley, the Tierra Mayor Valley, and the Beagle Channel (Fig. 2.3, Rabassa et al. 2000). Based on the morphology of the morainic bodies, Meglioli (1992) defined two different glacial advances that may be equated with the "Primera Angostura" and "Segunda Angostura" glaciations of the Strait of Magellan region. These glaciations can also be identified along the south polarization axis. Likewise, several glaciations have been recognized in the northern part of the island (Meglioli et al. 1990) with at least two along the Beagle Channel.

In the case of the Fagnano Lake Valley, it was occupied by a glacier from the Darwin Range that flowed eastward with more than 50 tributary glaciers. This generated an alpine-type landscape in the western region of the lake with ridges, cirques, truncated spurs, and hanging valleys. In contrast, a piedmont-type landscape developed in the eastern region with lateral moraines, glaciofluvial and glaciolacustrine terraces and landforms resulting from ice disintegration (Coronato et al. 2009). The Fagnano paleoglacier flowed through the center of Tierra del Fuego through

Fig. 2.3 Location and extent of major ice lobes along the center of Tierra del Fuego

three different lobes along the present-day valleys of the Fuego, Ewan, and San Pablo Rivers. These acted as drainage lines to the Atlantic Ocean with an estimated ice-covered area of 4000 km^2 (Fig. 2.3, Coronato et al. 2009).

The geological structures of the Fuegian Andes have preferably controlled the formation of glaciers of different sizes and thicknesses, which in turn form an extensive ice network. The glacier in Fagnano Lake has been defined by Coronato et al. (2009) as an outlet or discharge glacier, fed by smaller alpine glaciers in its upper and middle sections and with the appearance of a piedmont glacier in its lower portion. This differential ice behavior in the various sections has forced the onset of a mixed glacial landscape, both mountain glaciation and lowland glaciation.

The 14C dating of basal peatlands in sectors near Fagnano Lake shows that most of the area, especially the easternmost part and the S coast, was ice-free 12,300 years B.P. ago (Coronato et al. 2009). Fossil peat contained in the lower basal till deposits has provided 14C dates of 31,000–48,200 year B.P., indicating that a glacial advance occurred before the LGM (culminating ca. 25,000–23,000 cal. Years B.P.) (Coronato et al. 2009), perhaps in Isotopic Stage 4 (IS4), or else correspond to warmer episodes during the interstadial Isotopic Stage 3 (IS3, Sanci et al. 2021).

Previews works have mapped the terminal moraines surrounding Fagnano Lake (Nordenskjöld 1899; Quensel 1912; Bonarelli 1917). These authors have interpreted a bifurcated glacial lobe, one to the NE along the present San Pablo River and the other to the SE, toward Sloggett Bay and the Beagle Channel. Subsequently, Caldenius (1932) has mapped four glacial boundaries from the most recent to the oldest, named "Finiglacial", "Daniglacial", "Gotiglacial" and "Initioglacial". Although these four glacial advances existed, they do not correspond to the typical chronology of the Late Pleistocene in the Northern Hemisphere. Rather, these Patagonian glaciations correspond to events separated in time and extending throughout the Pleistocene (Coronato et al. 2004; Rabassa 2008). The first was recognized in the morainic ridges surrounding the head of Fagnano Lake and 20 km to the east, near Pescado Pond; the other three were located between the Lucio López, Noguera, and Irigoyen Hill mountain ranges, near the present Atlantic Ocean coast (Coronato et al. 2009). Meglioli (1992) has identified different drift units (i.e., glacigenic sedimentary deposits) around the lake and postulated the diffluence of a glacial lobe flowing northward along the northern shore. Three morainic arcs, composed of so-called Tills 5, 4, and 3, have been identified and interpreted as glacial events developed at Oxygen Isotopic Stages (OIS) 2, 6, and/or 10, respectively. Till 5 has also been recognized further east of the head of Fagnano Lake, coinciding with the location of the boundaries defined by Caldenius. Till 3 has been mapped even further to the E, in the San Pablo Valley (Meglioli 1992). Besides, it is believed that the moraines of the three oldest glaciations proposed by Caldenius (1932) represented the same phase.

Bujalesky et al. (1997) have described the glaciofluvial and glaciolacustrine sequence of a proglacial deltaic environment on the southern margin of Fagnano Lake. Fossil peat interbedded in lacustrine layers, covering the beds below and underlying the uppermost till unit, has given 14C dating ages between 39,500 and >58,000 years B.P. This would indicate, as previously mentioned, that the formation of the deltaic deposits would have taken place during a glacial advance before the LGM, probably in Isotopic Stage 4 (around 90–60 ka, in the early part of the Late Pleistocene). Coronato et al. (2002) have mapped glacial landforms in the eastern region of the lake headwaters and have suggested the presence of ground moraines east of Pescado Pond (Fig. 2.4). In this way, the ideas of the maximum ice expansion to the E proposed by Caldenius (1932) as well as those of Meglioli (1992) are modified.

2.3 Landforms and Landscape Associated with Neotectonic Origin

The geomorphological character of the study area and its origin are due to various episodes of tectonic and glacial modeling. The region corresponds to a transition area with two main units: one elevated and the other lower elevation areas. The elevated unit is represented by the mountain ranges, hill, and low rocky hills (300 and 250 m a.s.l. to the east in the Udaeta sector), while the lower elevation unit extends from the foothills to the north of the province.

Within the elevated unit different Fuegian Andes mountain ranges are recognized, such as the Aklekoyen Hill (500–600 m a.s.l.), the Knokeke Hill (998 m a.s.l.), the Jeujepen Mount (704 m a.s.l.), and the Irigoyen Hill (500 m a.s.l.) (Fig. 2.1b). In the lower elevation unit are moraines, drumlins, kettle-holes, glaciofluvial fans, kame

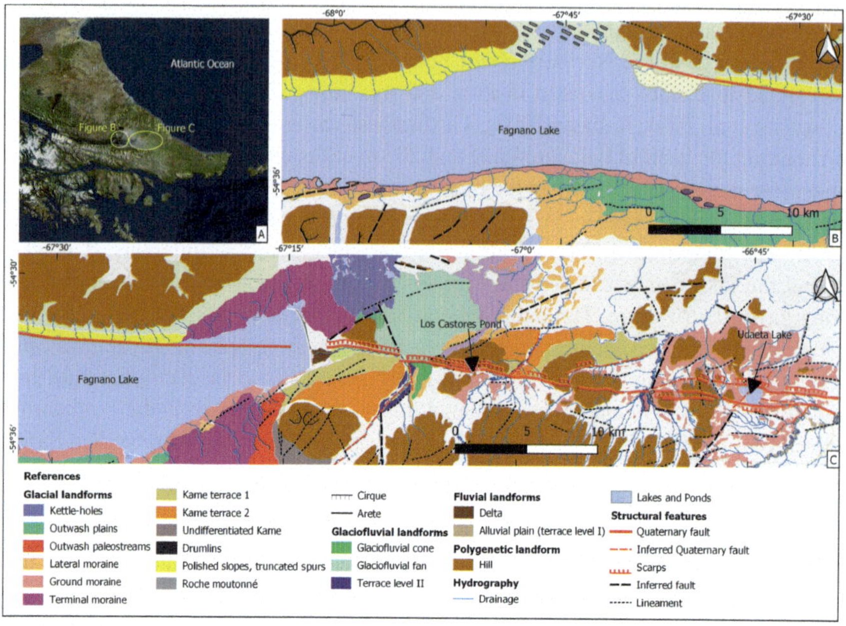

Fig. 2.4 Geomorphological map of the study area

terraces, outwash plains, glaciofluvial terrace levels and plains, peatlands, lakes and ponds (Fig. 2.4).

Toward the southern shore of Fagnano Lake, extensive landforms corresponding to ground moraines are recognized as part of the Frontal Complex (Fig. 2.4, Coronato et al. 2002). These ground moraine deposits are found attached to sandstone or forming strongly dissected plains and are exposed in the cliffs. In some sectors, there are intercalations with glaciofluvial levels and deposits corresponding to a proglacial delta (Bujalesky et al. 1997). The drainage network flows through outwash plains crossed by main rivers such as the Pescado, San Pablo, Ginebra, Láinez, and Irigoyen Rivers (Figs. 2.1, 2.2, and 2.5). In general, the morainic units are separated by elongated depressions, interconnected, and occupied by extensive peatlands.

The Frontal Complex includes different landforms that have been identified on the northern and southern margins of the eastern headwaters of Fagnano Lake, from lake level at 22 m a.s.l. to a maximum elevation of 117 m a.s.l. The complex presents spatial continuity with a northeast orientation and is dissected by the Turbio River Valley (Figs. 2.2 and 2.5), near its deltaic mouth at the Fagnano Lake headwaters. The complex includes moraine deposits with outcrops of basal till, outwash plains composed of massive clayey silts, and a frontal moraine composed of very compact yellowish-brown gravelly till, with abundant clasts, kames terraces that laterally limit and/or overlap the previous landforms.

The diversion of the Turbio River Channel is notorious evidence of drainage control caused by scarps generated by the MFFS activity (Fig. 2.5b).

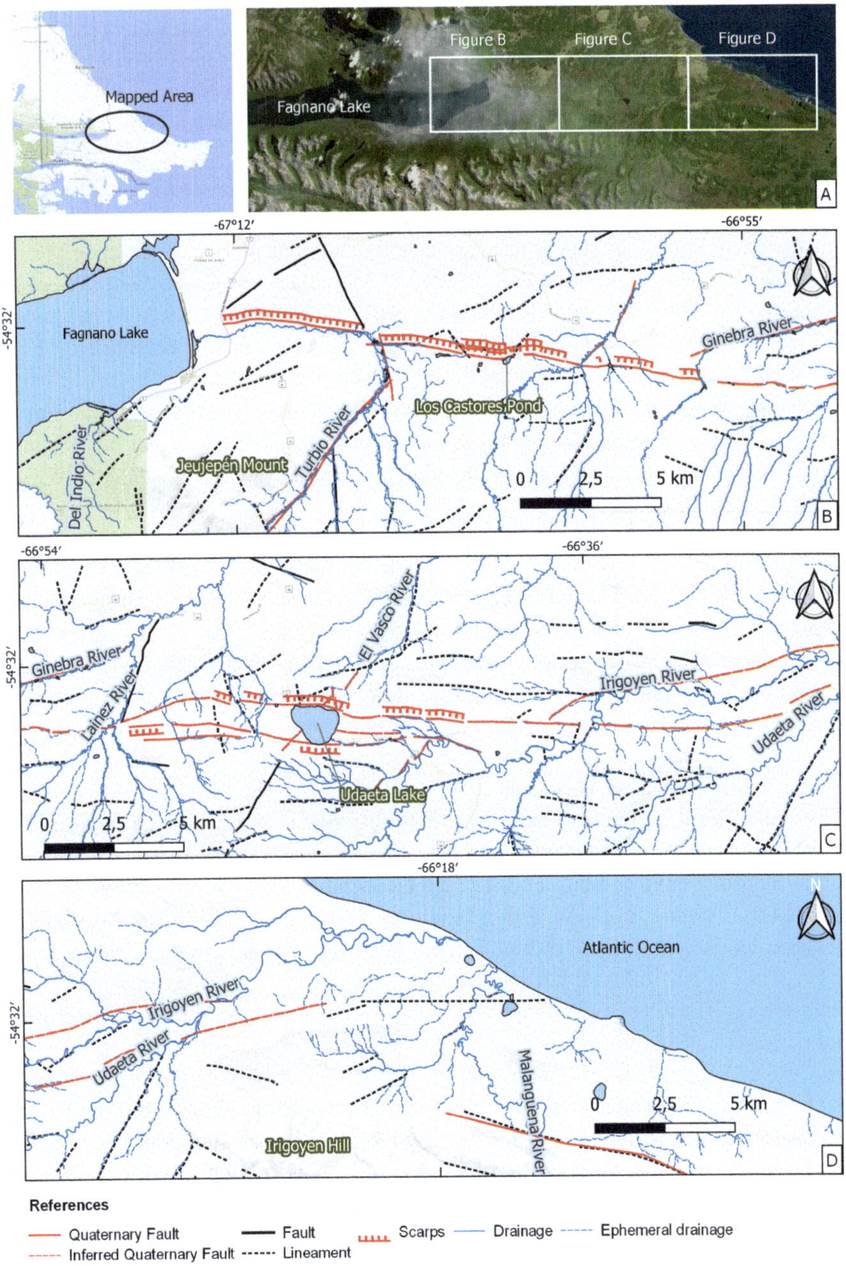

Fig. 2.5 Drainage and structural features

To the northeast of the Fagnano Lake headwaters, kettle-holes are recognized among the frontal glacifluvial deposits of the complex, the ground moraines on the slopes, and the foothills of the Aklekoyen Hill (Fig. 2.1b). Both the terraced elevations and the depressions are formed by glacifluvial gravel deposits, in some cases with outcropping thicknesses of 3 m.

The kame terraces are arranged in staggered levels between 287 and 179 m a.s.l. The highest terraced levels—between 287 and 251 m a.s.l.—are separated, forming a disconnected unit. The remnants are attached to the rocky slope of Jeujepen Mount and to the isolated rocky hills that border the Magallanes-Fagnano fault system to the north. The lower terrace is located between 179 and 228 m a.s.l. in a continuous and stepped form, attached to the upper terrace. These units are presented as "East Kame Terrace I" and "II", respectively (Figs. 2.4 and 2.6b).

Lateral moraines can be identified between the Turbio River and the San Pablo River, separated by elongated depressions that functioned as meltwater outlet channels. The moraines located south of the plate edge reach 240 m a.s.l. and are divided by valleys with recognized terraced levels.

The glaciofluvial fan described by Coronato et al. (2002) occupies the central part of the depression between the hill, bounded to the west by the Frontal Complex kames and rocky outcrops and to the east by ground moraines and kame terraces. The apex rises above the kame terraces and indicates provenance from one of the transverse valleys of the Lucio López Hill to the south. The distal portion interdigitates to the northwest with both the kettle-hole system and the outwash plain on which it rests to the northeast, and is blocked by ground moraines attached to the rocky hill slopes. It is composed of terrace-like or gently sloping gravel accumulations to the northeast, separated from each other by elongated, wide, flat-bedded channels of irregular design, thereby generating a partially functional drainage network of a distributive type. Thus interpreted, the fan is the result of glaciofluvial deposition in an ice-marginal environment. Four terrace levels have been identified in this sector (Fig. 2.6b). Terrace level IV, with a height of 220 m a.s.l., corresponds to lateral moraine deposits and is the oldest. Terrace level III, with a height of 200 m a.s.l., corresponds to glaciofluvial fan apex deposits and is dissected by the plate boundary. Terrace level II, with a height of approximately 145 m a.s.l., corresponds to glaciofluvial deposits and is arranged within the fluvial valley of the Turbio River, near the fan apex area. Terrace level I, with a height of 122 m a.s.l., corresponds to the current level of the fluvial valley. Onorato et al. (2019a) present a three-stage model for the evolution of this glaciofluvial fan, estimating that its development began to be deposited 26,000 + 4,500 B.P The formation of the glaciofluvial fan during the late Pleistocene may have taken place in a period of low MFFS activity. It is inferred that the MFFS was reactivated in the late Pleistocene-Holocene, generating scarps that dissected the glaciofluvial fan and caused the change in the flow direction of the Turbio River (Fig. 2.7). Given the important structural control in the design of the Turbio River Channel, these features in the landscape constitute some of the most remarkable tectonic evidence. For instance, the scarps recognized in this sector are a recent feature of seismic activity, as the 0.50–1 m scarp that would have formed with the 1949 earthquake (Costa et al. 2006).

Fig. 2.6 **a** North panoramic view apex levels of the glacifluvial fan attached to the slope of Mesa Hill; in a third plane the outwash plains are observed; **b** East view of recognized terraced levels in the Turbio River sector

Fig. 2.7 Eastern view of a linear valley in the Turbio River area. The MFFS is indicated with a red arrow

With the MFFS as the main axis, peatlands have been identified as developing over ancient glacial melt channels to the east of the Turbio River. Moraines up to 250 m a.s.l. have also been recognized, which are defined in this work as bottom moraines, mainly due to their lithological composition (basal till). These moraines are found attached to the sidewalls of hills with rocky cores that have been observed along the entire plate boundary and that lose height toward the east. Some of these forms correspond to the so-called whale back, as they exhibit an elongated shape parallel to the main axis of the structure (Figs. 2.7 and 2.8). These hills with rocky cores have an average height of 200 m, rounded summits, and are separated by valleys occupied by peatlands. In some cases, the summits show the development of wind gaps.

It is interpreted that this landscape has been affected by glacial advances throughout the Pleistocene. According to Meglioli (1992), these valleys occupied by peatlands are the result of the modeling of glacial meltwater streams active during the different glacial events of the Middle Pleistocene and/or during the Last Glaciation. Although the author has not recognized the presence of moraines in the region, he has mapped the extent of glaciofluvial gravel and sand deposits covering most of the rocky outcrops. He has called these deposits "Fuegian Gravels" and has interpreted them as glaciofluvial remnants of the Great Patagonian Glaciation (1.15–1.05 Ma, Coronato et al. 2004).

In the Ginebra River sector and to the east (Fig. 2.2b, c), the drainage has a design parallel to the main axis of the plate boundary. This drainage design is influenced by the fault system structures (Fig. 2.9). The southern block of the plate

Fig. 2.8 Northwest oblique aerial view of the whale back, scarp of fault and Los Castores sag pond. The MFFS is indicated with a red arrow

boundary presents mainly a parallel design, drainage inversion, channel diversion and displacement, and funnel design.

Sag ponds are quite frequent along the plate boundary (Figs. 2.2 and 2.8). Located both on the northern block and southern blocks, these features indicate an alternation

Fig. 2.9 Northwest oblique aerial view in the Ginebra River. The MFFS is indicated with red arrows

between up and down blocks. Los Castores Pond is interpreted as a sag pond generated on the southern block at the foot of composite scarps (Fig. 2.10, see Onorato et al. 2021). Further east, Udaeta Lake is recognized, which is considered another prominent evidence of recent tectonic activity. Regarding the tectonic origin of Udaeta Lake (Figs. 2.2c and 2.4), Onorato et al. (2017) suggest that it is a pull-apart-type traction basin due to its sub-circular shape, the morphology of the north and south shores structurally controlled by E-W trending scarps, the high slope gradient values in the basin, the subaqueous topography and the lake bottom sediments, among other characteristics. In addition, Onorato et al. (2019b) have applied geophysical methods that allowed them to identify these structures in-depth and have interpreted them as faults on the north and south shores, in agreement with the scarps observed at the surface.

Morainic deposits, glacial melt valleys covered by peatlands, and hills with rocky cores have been recognized on the periphery of Udaeta Lake. The moraines are located between 200 and 235 m a.s.l. (Fig. 2.11). They are composed of basal till and separated by valleys covered by peatlands.

The northeast-flowing fluvial valleys of the Láinez and El Vasco Rivers stand out in the landscape (Fig. 2.2c). The Irigoyen River flows in sections to the east almost parallel to the course of the MFFS main structures affecting the plate boundary. Before reaching the Atlantic coast, these last sections of the fault system generate

Fig. 2.10 North view of Los Castores pond. The MFFS is indicated with a red arrows view of the sag ponds at the foot of the scarp. See location in Figs. 2.1b and 2.2b

Fig. 2.11 **a** Southwest view of the scarp of the Udaeta Lake, the location of the described profile by Onorato et al. (2016); **b** Photographs of the profile

small hills interpreted as pop-up, which are observed in some cases associated with fault scarps (Onorato et al. 2021).

2.4 Final Comments

The present chapter resumes the first investigations to determine the influence of neotectonic activity in a region affected by alpine and continental engulfment during the Quaternary along the South American and Scotia plate boundary. This work provides morphotectonic characteristics and evidence of neotectonisms both in landforms.

The MFFS plays a key role in the genesis and/or modification of glacial and fluvioglacial landforms, such as morainic deposits, glacifluvial terraces, rivers and lake basins, as well as in their tectonic-geomorphological evolution.

Morphotectonic evidence of the neotectonic activity of the MFFS indicates a sinistral horizontal main displacement, such as fault scarps alternating their free face to the north and south, linear valleys and drainages, evidence of drainage control, such as offset drainages, wind gaps, and drainage inversion, sag ponds and pull-aparts, linear and pressure fault ridges, among others. The drainage has a strong structural control in some sectors, presenting a design parallel to the main axis of the plate boundary. The rivers that stand out are the Láinez and El Vasco Rivers, which flow northeastward, and the Irigoyen River, which in lower sections flows eastward almost parallel to the course of the main structures of the MFFS, while in upper sections flows with a S–N course.

This work indicates the characteristics of the landforms that result from tectonic activity and its influence on the geomorphological configuration along the plate boundary. The geomorphological influence of structures associated with the MFFS and the tectonic origin of nearby water bodies, such as the Los Castores Pond, are open to discussion. The possibility of obtaining geochronological ages to constrain landforms, sedimentary units, and tectonic activity is crucial for the paleoseismic understanding of this section of the MFFS.

References

Audemard FA (2003) Estudios paleosísmicos por trincheras en Venezuela: métodos, alcances, aplicaciones, limitaciones y perspectivas. Revista Geografica Venezolana 44:11–46

Bonarelli G (1917) Tierra del Fuego y sus Turberas. Ministerio De Agricultura De La Nación Sección Geología, Mineralogía y Minería XII 3:1–119

Borrello AV (1972) Cordillera Fueguina, Academia Nacional de Ciencias, Cordoba, Argentina, pp 740–753

Bujalesky GG, Heusser CJ, Coronato AM, Roig CE, Rabassa JO (1997) Pleistocene glaciolacustrine sedimentation at Lago Fagnano, Andes of Tierra del Fuego, southernmost South America. Quat Sci Rev 16:767–778

Caldenius CC (1932) Las glaciaciones cuaternarias en la Patagonia y Tierra del Fuego. Dirección General De Minas y Geología 95:1–164

Coronato A, Mazzoni E, Vázquez M, Coronato F (2017) Patagonia: Una síntesis de su geografía física. Universidad Nacional De La Patagonia Austral 1:218

Coronato A, Roig CE, Mir X (2002) Geoformas glaciarias de la región oriental del Lago Fagnano, Tierra del Fuego, Argentina. In: Cabaleri N, Cingolani C, Linares E, López de Luchi M, Ostera

H, Panarello H (eds) Actas del XV Congreso Geológico Argentino, Tomo II, pp 457–462. CD-Rom. Artícle Nº 24, 6 pp

Coronato A, Meglioli A, Rabassa JO (2004b) Glaciations in the Magellan Straits and Tierra del Fuego, Southernmost South America. In: Ehlers J, Gibbard P (eds) Quaternary glaciations—Extent and chronology. Part III. quaternary book series. Elsevier, pp 45–48

Coronato A, Seppälä M, Ponce JF, Rabassa J (2009) Glacial geomorphology of the Pleistocene Lake Fagnano ice lobe, Tierra del Fuego, southern South America. Geomorphology 112:67–81

Costa CH, Smalley R, Schwartz DP, Stenner HD, Ellis M, Ahumada EAY, Velasco MS (2006) Paleoseismic observations of an onshore transform boundary: the Magallanes-Fagnano fault, Tierra del Fuego, Argentina. Revista de la Asociacion Geológica Argentina 61:647–657

Forshyth DW (1975) Fault plane solutions and tectonics of the South Atlantic and Scotia Sea. J Geophys Res 80:1429–1443

Isla F, Schnack E (1995) Submerged moraines offshore northern Tierra del Fuego, Argentina. Quat South Am Antarct Penins 9:205–222

Klepeis KA (1994) The Magallanes and Deseado fault zones: major segments of the South American-Scotia transform plate boundary in southernmost South America, Tierra del Fuego. J Geophys Res: Solid Earth 99:22001–22014

Lodolo E, Menichetti M, Bartole R, Ben-Avraham Z, Tassone A, Lippai H (2003) Magallanes-Fagnano continental transform fault (Tierra del Fuego, southernmost South America). Tectonics 22

Meglioli A, Evenson E, Zeitler P, Rabassa JO (1990) Cronología relativa absoluta de los depósitos glaciarios de Tierra del Fuego, Argentina y Chile. In: XI Congreso Geológico Argentino, San Juan. pp 457–460

Meglioli A (1992) Glacial Geology of southernmost Patagonia, the Strait of Magellan and northern Tierra del Fuego. Lehigh University, Bethlehem, U.S.A

Musotto LL, Borromei MA, Bianchinotti MV, Coronato A, Menounos B, Osborn G, Marret R (2017) Postglacial environments in the southern coast of Lago Fagnano, central Tierra del Fuego, Argentina, based on pollen and fungal microfossils analyses. Rev Palaeobot Palynol 238:43–54

Nordenskjöld O (1899) Geological map of the Magellan territories with explanatory notes. Kungl. Boktryckeriet. PA Norstedt & Söner

Olivero EB, Malumián N (2008) Mesozoic-Cenozoic stratigraphy of the Fuegian Andes, Argentina. Geol Acta 6:5–18

Onorato MR, Perucca LP, Coronato A, Rabassa JO, López R (2016) Seismically-induced soft-sediment deformation structures associated with the Magallanes-Fagnano Fault System (Isla Grande de Tierra del Fuego, Argentina). Sed Geol 344:135–144

Onorato MR, Coronato A, Perucca LP, Rabassa JO, Lopez R (2017) Morpho-bathymetry and surficial morphology of Udaeta Lake, along the Magallanes-Fagnano fault system, Tierra del Fuego, Argentina. J S Am Earth Sci 76:1–10

Onorato MR, Perucca LP, Coronato A, Lopez R, Blanc PA (2019a) Evidencias morfotectónicas en el sistema de fallas Magallanes-Fagnano, borde transformante entre las placas Sudamericana y Scotia, Isla Grande de Tierra del Fuego, Argentina. Revista De La Asociación Geológica Argentina 77(1):47–61

Onorato MR, Prezzi C, Orgeira MJ, Perucca LP, Coronato A, Lopez R, Magneres I (2019b) Geophysical characterization of Udaeta Lake as a possible pull-apart basin associated to quaternary tectonic activity along Magallanes-Fagnano fault system. Quat Int 523:67–79

Onorato MR, Perucca LP, Coronato A, Prezzi C, Blanc PA, López R, Magneres I (2021) Morpho-tectonic characterization along the eastern portion of the main trace of Magallanes-Fagnano fault system in Tierra del Fuego, Argentina. J S Am Earth Sci 112:103550

Pelayo AM, Wiens D (1989) Seismotectonics and relative plate motions in the Scotia Sea region. J Geophys Res 94:7293

Quensel P (1912) Die Geologie der Juan Fernandezinseln. Bull Geol Inst Upps 11:252–290

Rabassa J (2008) Late Cenozoic glaciations in Patagonia and Tierra del Fuego. In: Developments in quaternary sciences. pp 151–204

Rabassa J, Coronato A, Bujalesky G, Salemme M, Roig C, Meglioli A, Heusser C, Gordillo S, Roig F, Borromei A, Quattrocchio M (2000) Quaternary of Tierra del Fuego, Southernmost South America: an updated review. Quat Int 68:217–240

Roy S, Vassallo R, Martinod J, Ghiglione MC, Sue C, Allemand P (2019) Co-seismic deformation and post - glacial slip rate along the Magallanes-Fagnano fault, Tierra del Fuego, Argentina. Terra Nova 1–10

Sanci R, Orgeira MJ, Coronato A, Tófalo R, Perdomo P, Gogorza C, Quiroga D, López R, Palermo P (2021) Late Pleistocene glaciolacustrine MIS 3 record at Fagnano Lake, Central Tierra del Fuego, southern Argentina. Quat Res 102:53–67. https://doi.org/10.1017/qua.2020.93

Sandoval FB, Pascale GP De (2020) Slip rates along the narrow Magallanes fault system, Tierra Del Fuego region, Patagonia. Sci Rep 10(1):8180

Smalley R Jr, Kendrick E, Bevis M, Dalziel I, Taylor F, Lauría E, Barriga R, Casassa G, Olivero E, Piana F (2003) Geodetic determination of relative plate motion and crustal deformation across the Scotia-South America plate boundary in eastern Tierra del Fuego. Geochem Geophys Geosyst 4(9):1070. https://doi.org/10.1029/2002GC000446

Torres Carbonell PJ, Rodríguez Arias L, Atencio MR (2017) Geometry and kinematics of the Fuegian thrust-fold belt, southernmost Andes. Tectonics 36:33–50

Vedder JG, Wallace RE (1970) Map showing recently active breaks along the San Andreas and related faults between Cholame Valley and Tejon Pass, California (No 68-311). US Geological Survey

Winslow MA (1982) The structural evolution of the Magallanes Basin and neotectonics in the southernmost Andes. In: Craddock C (ed) Antarctic geoscience. University of Wisconsin Press, Madison, pp 143–154

Chapter 3
Glacial and Periglacial Geomorphological Study Associated with the Evolution of Rock Glaciers in the High Basin of the Río Cochiguás, Coquimbo, Chile

Catalina Pino-Rivas

Abstract In the high mountain range of the Elqui Province, Chile, a landscape dominated by glacial and periglacial processes is recognizable. Glaciers have been identified as potential water reservoirs, although their dynamics and hydrological contribution have not been fully understood. Therefore, the study of these landforms becomes relevant to generate sustainable policies for the management of these potential water resources. Here, the volumetric variation of two rock glaciers in the basin of the Río Cochiguás for the period 1956–2018 is quantified, in order to establish the influence of geomorphological evolution, climatic variations and their possible hydrological implications to the watershed. A geomorphological map of the area was made. Topographic data were taken using differential GPS and a characterization of their detrital cover was developed. With this, it was possible to compare the volumes of the rock glaciers over time by comparing the current topography with surfaces reconstructed from aerial photographs of 1956 and 2000 of the same area. Glacial and periglacial landforms, cryogenic features among other geomorphic features were identified through the mapping. Some glacial landforms exhibit evident cryogenic features such as selection by freezing, slight grooves and superficial ridges and furrows, and a steep lateral slope composed of finer material. Valley-bottom rock glaciers, protalus lobes, gelifluxion slopes and avalanche deposits were also identified. The debris cover of the rock glaciers was conditioned by the high degree of weathering of the source outcrops. In addition, the angle of the frontal slope of these landforms is greater than the angle of repose, and they exhibit some surface features as ridges and furrows. These features indicate that these bodies are active, and that physical weathering processes play an important role in their dynamics. The comparison of rock glacier volumes indicated considerable changes, given the decrease in precipitation and the increase in temperature in recent years in the area, which would have a direct impact on freshwater reserves for the region.

C. Pino-Rivas (✉)
BGC Ingeniería, Isidora Goyenechea 3520 piso 9, Las Condes, Santiago, Chile
e-mail: catapinorivas@gmail.com

A. Coronato and G. B. Alves (eds.), *Latin American Geomorphology*, The Latin American Studies Book Series, https://doi.org/10.1007/978-3-031-55178-9_3

Keywords Rock glaciers · Periglacial environment · Glacial environment · Cryogenic landforms · Glacial features

3.1 Introduction

In the Semi-arid Andes, specifically in the high basin of the Río Cochiguás in the Coquimbo region, Chile, the glacial and periglacial environment is composed of complex geomorphological system, in which some landforms as rock glaciers are significant in number and size. However, this is a recent research topic in Chile, where it is still uncertain the dynamics, the inner structure and the water supply to the basins of these ice bodies. Additionally, under climatic tendencies that point towards a gradual atmospheric warming and an evident diminution in the precipitation in the area, it is necessary to study the effects that these processes would have on this type of cryogenic landforms.

This mountain landscape is characterized by steep slopes and deep valleys. The higher portion of the basin is related to a glacial and periglacial environment with the occurrence of talus, linear slopes, rock glaciers and solifluction and gelifluction slopes. Likewise, an important presence of rock and snow avalanches in the area (Paskoff 1967; Brenning 2005).

Due to the scarcity of precipitation during most of the year, the water reserves in this sector are mostly depending upon the quantity of snow accumulated in the High Cordillera and its later melting and infiltration, which feeds the basins of the main fluvial processes, as it is the case of the Elqui, Limarí and Choapa Rivers in the Coquimbo region. This climatic configuration of high mountains and their consequent fluvial streams constitute a primordial resource in the development of agricultural and cattle rising of the zone (Janke et al. 2015).

Presently, the region is being affected by the phenomenon of the "Central Chile mega-draught" that translates into a water deficit of approximately 30%, between 2010 and 2015 (Garreaud et al. 2017), and largely overcoming any other event during the rainy historical record. This problem would become more relevant in the future due to the tendencies that indicate a diminution of the rainfall and an increase in the temperature of the air in the surface of the Central Andes of Chile, which would continue also along the twenty-first century.

All this has promoted the interest of the communities living near mountain areas and the Chilean Government in investigating the availability of water resources, its dynamics through time and its spatial distribution, in which the rock glaciers have been particularly noted as potential water resources in arid zones.

This chapter aims to contribute to the knowledge of the cryosphere in the Andes by studying the morphology of rock glaciers, and the characterization of these landforms in the field. In addition, trying to relate their present characteristics with their volumetric variation over time and their possible water implications in high mountain basins.

3.2 The Study Area

The study area corresponds to the sub-basin of the Río Cochiguás, part of the basin of the Río Elqui, located between the 30°20′ and 30°29′S (Fig. 3.1a). It is limited to the north and northeast by the basin of Río Ingaguás, and has its sources in the high Cordillera at the boundary with Argentina. From an administrative point of view, the sector is located in the Paihuano County, Province of Elqui in the Coquimbo region, 46 km to the southeast of the Alcohuaz community in the Elqui Valley.

The studied rock glaciers are located at the inner part of the Quebrada Caballos that also belongs to this basin, placed between 30,44° and 30,47° south latitude, at the foot of a rock slope that indicates the international boundary between Chile and Argentina, at the heads of the Río Cochiguás.

The study area is located in the South-Central Andes, at latitude 30°28′S, where the subduction angle between the Nazca and South American plates is about 10°, a sector that is usually referred to as the "Central Chilean flat subduction segment" (Barazangi and Isacks 1976; Jordan et al. 1983; Gutscher et al. 2000) and as the "Pampean flat subduction segment" by Folguera et al. (2002). This segment is characterized by an intense coupling between the Nazca and South American plates, a highly compressed continental crust and the absence of a longitudinal valley separating the Cordillera Frontal (morphostructure of the Andes Mountains) from the Cordillera de la Costa (Pardo et al. 2006). Yáñez et al. (2001) suggest that the planar subduction in this region is related to the subduction of the Juan Fernández aseismic ridge at 33°S,

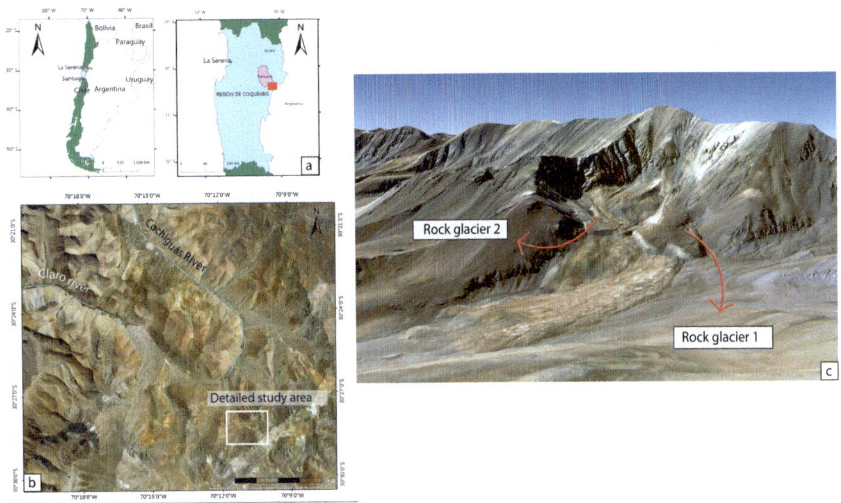

Fig. 3.1 Study area. **a** Regional study area. The red square indicates the location of the detailed study area in (**b**). **b** Detailed study area showing the main rivers that originate in the basin. The black square indicates the location of the two main rock glaciers of this study. **c** Rock glaciers 1 and 2 studied in this chapter

which has been continuously subducting beneath South America at the same point since 10 Ma ago.

The large-scale geomorphology in the study area is characterized by a marked increase in elevation along the west–east transect (Aguilar et al. 2011, 2013). This first-order geomorphological feature represents a topographic front that separates two elongated morphostructural units in a north–south direction, corresponding to the Coastal Cordillera to the west and the Frontal Cordillera to the east.

The Coastal Range is characterized by a series of coastal platforms that present low slope (Paskoff 1970; Ota et al. 1995; Rodríguez et al. 2015), while, to the east, this reaches its maximum elevation at 3,200 m a.s.l. This morphostructure corresponds mainly to an east-vergent homocline of Triassic-Lower Cretaceous volcano-sedimentary rocks, unconformable to the metamorphic and sedimentary basement of Devonian-Permian age (Rivano and Sepúlveda 1991).

The Cordillera Frontal, on the other hand, reaches elevations of up to 6,700 m a.s.l. It is formed by a core of Carboniferous-Permian magmatic units (Nasi et al. 1990; Pineda and Calderón 2008), which are covered to the west by a block of Triassic-Lower Cretaceous volcano-sedimentary rocks and intruded by a Late Cretaceous-Early Paleocene magmatic arc (Mpodozis and Cornejo 1988; Nasi et al. 1990; Pineda and Calderón 2008). To the east, on the other hand, the basement is intruded by a block composed mostly of Permo-Triassic volcanic and magmatic rocks overlain unconformably by Oligocene-Miocene volcano-sedimentary rocks (Maksaev et al. 1984; Nasi et al. 1990; Bissig 2001; Murillo et al. 2017).

Finally, the area east of the main topographic front corresponds to the Main Cordillera, which is defined by a core of Oligocene-Miocene volcano-sedimentary rocks (Charrier and Baeza 2002; Mpodozis et al. 2009; Jara and Charrier 2014). These rocks are covered in unconformity by Miocene subhorizontal volcanic rocks and intruded by a north–south-trending Miocene magmatic arc (Mpodozis et al. 2009; Jara and Charrier 2014).

Regarding the lithology of the sector (Fig. 3.2), a series of intrusive, metamorphic and sedimentary units outcrop in the Cochiguás River Sub-Basin, most of them with ages ranging from the Paleozoic to the Triassic (Mpodozis and Cornejo 1988). These rocks correspond to the basement of the geomorphological units detailed below and represent the main lithology of the different deposits and geoforms identified in the sector.

The main study area (Fig. 3.2) is essentially composed with Carboniferous rocks of the Elqui Super Unit (SUE), Permian-Triassic rocks of the Ingaguás Super Unit (SUI) and Miocene rocks of the Infiernillo Unit, all of them intrusive rocks.

On the one hand, the Elqui Super Unit is a group of plutons formed by leucocratic to mesocratic granitoids, coarse-grained, with synmagmatic, cataclastic and mylonitic banding, furrowed by swarms of basic (andesitic and andesitic-basaltic) and felsic dykes. It includes from gabbros to monzogranites.

In the study area, this super unit is represented by leucocratic granodiorites and monzogranites of biotite and muscovite, of medium to coarse grain and light grey to white colour (Cochiguás Unit); and by granodiorites and cataclastic granites of very

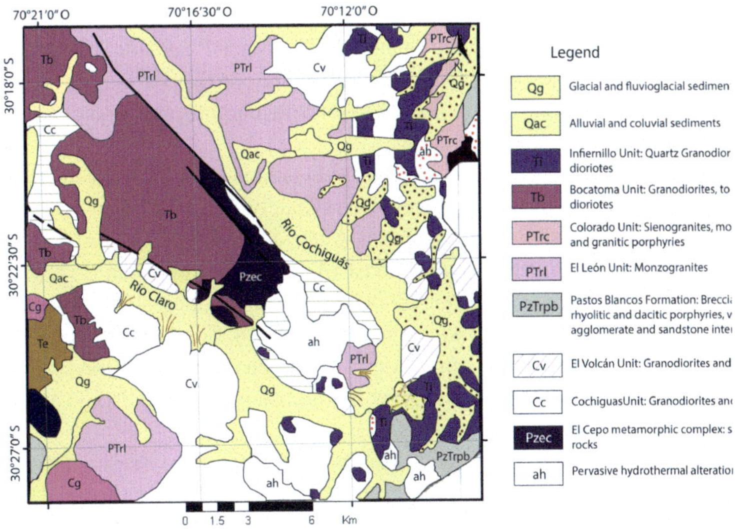

Fig. 3.2 General geological map of the area with the main geological units defined by Mpodozis and Cornejo (1988). Modified from Mpodozis and Cornejo (1988)

coarse grain and hypidiomorphic texture, affected by fracturing and recrystallization (El Volcán Unit).

In the lower course of the Cochiguás River, the Cochiguás Pluton, the largest of the plutons of the unit of the same name, is well exposed. In the area of the Claro River, the plutons are crossed by swarms of subvertical basic dykes, which constitute more than 50% of the total volume of the pluton in some outcrops. The El Volcán Unit, on the other hand, is located in the area near the border with Argentina, intruding the Hurtado Formation to the west and the Pastos Blancos Formation to the east. The latter marks the eastern limit of the Elqui Super Unit intrusives through an intrusive-tectonic contact.

The Ingaguás Super Unit, on the other hand, is an association of generally hololeucocratic plutons, which are mostly disposed to the east of the outcrops of the Elqui Super Unit. The granitoids of this super unit intrude both the schists of the El Cepo Metamorphic Complex and the Pastos Blancos Formation and the granitoids of the Elqui Super Unit.

In the central sector of the study area (Fig. 3.2), this super unit is represented by medium-grained "pink" leucocratic monzogranites and syenogranites of the El León Unit.

The last group of intrusive rocks is composed of the Infiernillo Unit, of Miocene age, which correspond to medium-fine-grained quartz diorites, granodiorites, tonalites and subordinate andesitic porphyries. The bedding rocks are the El Cepo Metamorphic Complex, the Pastos Blancos Formation, Paleozoic granitoids and the Doña Ana Formation. Rocks of this unit outcrop in the eastern sector of the study area, in a N–S strip near the border with Argentina.

In addition to the intrusive rocks mentioned above, there is an important area to the south-southeast of the study area made up of Upper Paleozoic-Lower Triassic volcanic rocks of the Pastos Blancos Formation, which correspond mainly to rhyolitic lavas, ignimbritic flows, tuffs, intercalations of volcanic breccias and scarce levels of sandstones and conglomerates.

About the climate, in general terms, the study area is located in a transition zone between arid and semi-humid climates, immediately south of what is known as the "Arid Diagonal", which crosses South America from the Gulf of Guayaquil, in Ecuador, and ends on the Atlantic coast of Patagonia, between 25° and 27°S (Eriksen et al. 1983; Zech et al. 2006; Ginot et al. 2006). To the north of this diagonal there is a greater influence of the Amazon System (SMS, South American Monsoon System), while to the south it is conditioned by the Pacific Circulation System and the westerlies, where glaciers generated by humid fronts from the Pacific and winter precipitation develop (Ammann et al. 2001).

The climate of the area also depends on the presence of the Pacific anticyclone. This is a high-pressure system located in the southeastern Pacific Ocean where the atmospheric pressure is higher than the surrounding area, which produces a downward movement of the air masses that inhibits the development of cloudiness and precipitation, and favours clear skies with high solar radiation (Escobar and Aceituno 1998).

In general, the current climate of the study area according to the classification of W. Köppen (in Errázuriz et al. 1998) corresponds to tundra due to the effect of altitude with little or no precipitation. This climate is a subtype of the high-altitude marginal desert climate and develops continuously in the highest part of the Andes Mountains, from 28°S.

Regarding seasonal variations, the area is characterized by cold winters and dry summers (Fiebig-Wittmaack et al. 2012), by a strong diurnal temperature variation between day and night, and a strong altitudinal effect on temperature (Azócar 2013). Most of the area's humidity is due to solid precipitation (snow) between May and August (Gascoin et al. 2010). However, during the summer, small snowfalls caused by humid air masses from the eastern side of the Andes are observed, mainly between January and March (Garreaud and Rutllant 1997) and also by convection of humid air from the west in the afternoons.

The 0 °C Mean Annual Air Temperature (MAAT) isotherm is located at approximately 4,150 m a.s.l. However, it tends to be at 3,700 m during the coldest month (July) and over 4,800 m in the hottest month (January) (Azócar 2013).

On the other hand, in the highest zones of the study area, and in particular where rock glaciers 1 and 2 are located, the MAAT varies between 0 °C (at 4,150 m) and −2° (at 4,400 m), according to the calculation developed by Azócar (2013) (Fig. 3.3).

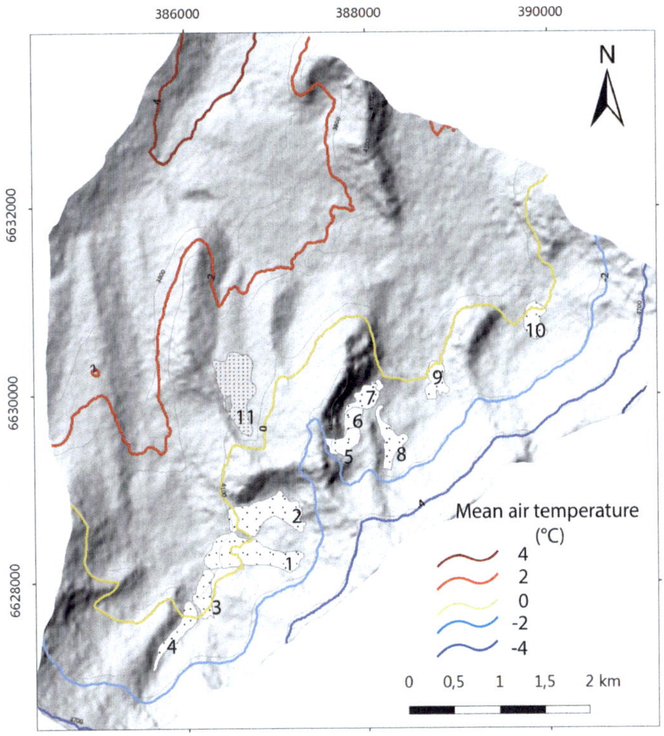

Fig. 3.3 Annual mean air temperature curves in the study area. Schematic developed from the MAAT modelling of Azócar (2013). UTM 19S coordinates

3.3 Periglacial Environment and Rock Glaciers

To understand what these cryo-landforms are, it is necessary to understand the context under which they have developed. According to the Terminological Guidebook of the South American Cryology (Trombotto et al. 2014), the periglacial environment is a cold and cryogenic environment with the occurrence of permafrost in depth, or permanently frozen soils, and the possible presence of underground ice trapped and preserved under natural conditions after a long time. Moreover, the freezing processes dominate with cycles of freezing and melting that affect the rocks and the upper portion of the soil, as well as solifluction, gelifluction and other cryogenic processes that model the geomorphology of the Andean regions.

The cryogenic rock glaciers are the expression of mountain permafrost (Barsch 1996) that correspond to landforms of periglacial genesis. The common definition corresponds to located bodies (in the slopes) or tongue-like (at the inner portion of the valleys) composed of unconsolidated materials, which are permanently frozen, supersaturated of interstitial ice, and which moves downslope by creeping as a consequence of the deformation of the ice contained inside (Barsch 1996).

There exists presently a debate about the periglacial or glacial origin of a rock glacier, which is closely related with its inner structure, a matrix of ice and rock or an ice core (Janke 2013). Those authors defending the periglacial model (cryogenic rock glaciers) believe that the interstitial cemented ice and/or the segregated ice (ice lenses) provokes a lithostatic pressure that allows the glacier movement by means of creeping (Wahrhaftig and Cox 1959; Barsch 1978, 1996; Wayne 1981; Haeberli 1985). Besides, those favouring a glacial model (glacigenic rock glaciers) suggest that an ice core would have its origin in an uncovered glacier (or "white"), covered and buried by debris (Potter 1972; Whalley 1974; Whalley and Azizi 1994; Potter et al. 1998). In this sense, the rock glaciers are frequently considered as part of a continuum in the landscape, that is, a cycle that describes the transition between uncovered glaciers, rock glaciers and slope deposits (Corte et al. 1987; Johnson 1984; Giardino and Vitek 1988). However, the present evidence support both models (Janke 2013).

It is accepted today that a rock glacier may have a cryogenic origin or a glacigenic origin.

Humlum (1988) developed a classification of debris ice components in mountainous areas based upon morphological criteria. In this classification, many landforms are recognized, such as valley glaciers, ice caps, cirque glaciers, rock glaciers (tongue shaped, valley floor, piedmont or spatulate), protalus rampart and extant or relict snow patches.

The degree of activity of a rock glacier is one of the most used classifications, proposed by Barsch (1996), who, starting from a set of dynamic characteristics, performed interpretations according to the presence of ice to the inner part of rock glacier, based upon the principle of the presence of sub-superficial ice in different cut sections (Haeberli et al. 2006). In this category, the rock glaciers are classified into active landforms, that means, in movement and with ice in the inner side, inactive landforms, without movement, but still with ice in its interior, and as fossil or relict, without movement and where the ice content has melted completely.

According to Barsch (1996), a rock glacier is considered as active when it shows signs or evidence of movement, when its front is strongly sloping (around 30–35°) or when in a vertical section it exhibits sedimentation processes and visible signs of rock fall. Their active surfaces present superimposed lobes, undulations and grooves characteristically caused by movement.

Besides, an inactive rock glacier is one that has stopped moving and it shows evidence of past but not recent movements. While its front presents a slope smaller than 30º, its surface is chaotic with depressions and collapse signals. The inactivity is generally the result of warming tendencies that have caused a diminution of the ice content of the ground (Trombotto et al. 2014). Both active and inactive forms are commonly grouped as intact rock glaciers, due to the difficulty in differentiation between active and inactive landforms by means of remote sensing (Barsch 1996).

Finally, a relic or fossil rock glacier is a mass of rock fragments and finer materials, in a certain slope, that shows evidence of the last movement, but it has lost all the soil and ground ice. Its surface is already vegetated and the fronts present slopes of less than 20° (Trombotto et al. 2014).

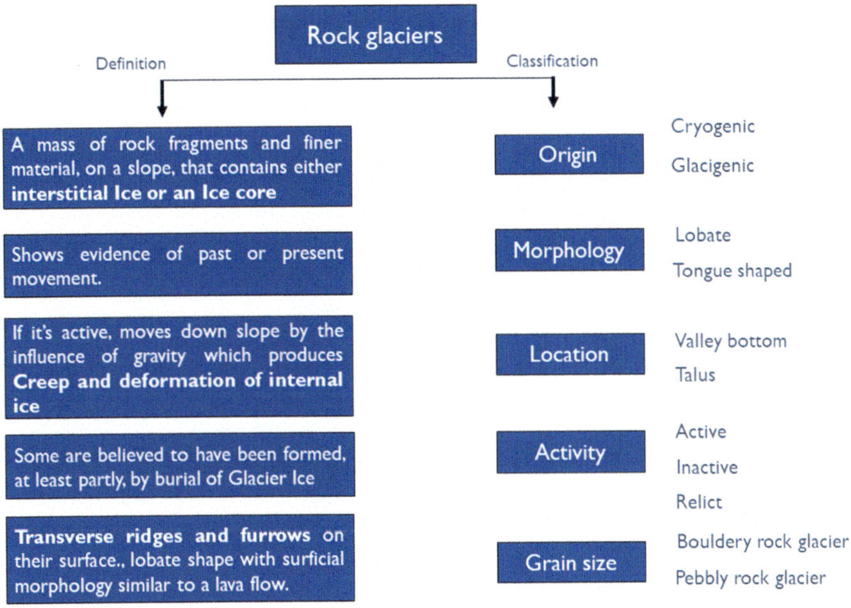

Fig. 3.4 Definition and general classification of rock glaciers based on origin, morphology, location, activity and size of blocks in their composition

Concerning the lithology, this is an important factor in the shape and size of the debris that composes the rock glacier. According to studies done in the Alps, the larger clasts are associated with crystalline rocks, for instance, igneous rocks, conglomerates and limestones, whereas those of smaller size are related to foliated and/or high porosity rocks, such as slates (Ikeda and Matsuoka 2006). This has allowed to classify the rock glaciers into two types according to the size and shape of the debris: the pebbly rock glaciers and the bouldery rock glaciers (Ikeda and Matsuoka 2006). Figure 3.4 summarizes and defines what rock glaciers are, how they are formed and their classification according to the literature reviewed in this chapter.

3.4 Methodology

A geomorphological characterization performing a remote mapping was done. High resolution of imagery (1.7 m) of February 2015 obtained from the National Service of Geology and Mining was used by means of the base map Worldview from Arcgis. The main morphological features of the area were identified, and two rock glaciers were chosen to be studied.

Afterwards, a geomorphological map at 1:25.000 scale was prepared based upon the information taken in the field, determining common criteria to the associated morphologies and characterizing the unconsolidated deposits of the area. The elaboration of this map was performed using the ArcGIS 10.4 software and the Adobe Illustrator 2018 software. With these digital tools a morphometric analysis of slopes and solar exposition was completed.

Likewise, a lithological and sedimentary characterization of the rock glaciers was performed with the objective of identifying the rocks' provenance and the patterns of compression, based upon the methodology used by Monnier et al. (2014). This methodology consists in the election of experimental points (sampling and measurements) within and around the landform, whose objective is to complement the surficial mapping and the delimitation of the main units.

To study the volumetric and morphological changes of the rock glaciers, the photogrammetric method was used. It allows the generation and comparison of Digital Elevation Models (DEM) constructed on different dates and, starting with them, it was possible to estimate the change of volume of the rock glaciers during a certain period ($\Delta h/\Delta t$). With this purpose, aerial photographs taken by the Instituto Geográfico Militar (IGM) of Chile were used as the basis of a Digital Elevation Model of high resolution (10 m) provided by the Servicio Nacional de Geología y Minería of Chile.

Control and elevation points through GNSS (GCPs, Ground Control Points) using two receptors of double frequency were taken to generate the DEM of the years 1955 and 1999, both in outcrops close to the study area, at the foot of the rock glaciers (Fig. 3.5). These control points have been taken to improve the precision of the generated elevation models, providing better precision. These points were chosen taking into consideration the conditions of the terrain such as climate, accessibility, logistics, as well as the proximity of rock glaciers.

Besides, the orthorectification of the aerial photographs HYCON 1955, GEOTEC 2000 and the generation of MDEs (each pair in a separate way) were generated using the software PCI GEOMATIC 2016. In this manner, the software provides a DEM of metric resolution and with a good level of detail, which may be improved by eliminating GCPs with high values of residuals in X, Y or in Z.

With the base of the results obtained from the calculation of the altimetric differences between the generated DEM, the behaviour of the rock glaciers was analysed for the period 1955–2018 corresponding to each difference, recognizing the main zones of loss or increase of volume and detecting the changes in thickness.

3.5 Geomorphological Context

As a first look, this section studies the geomorphological characteristics of the study area where the two rock glaciers of study are located in order to contextualize them, so as to understand their characteristics and historical variations.

Presently, the sub-basin of the Río Cochiguás is controlled by periglacial processes. This type of processes implies a readjusted and remodelled landforms and glacigenic deposits developed during a glaciation (Ballantyne 2002).

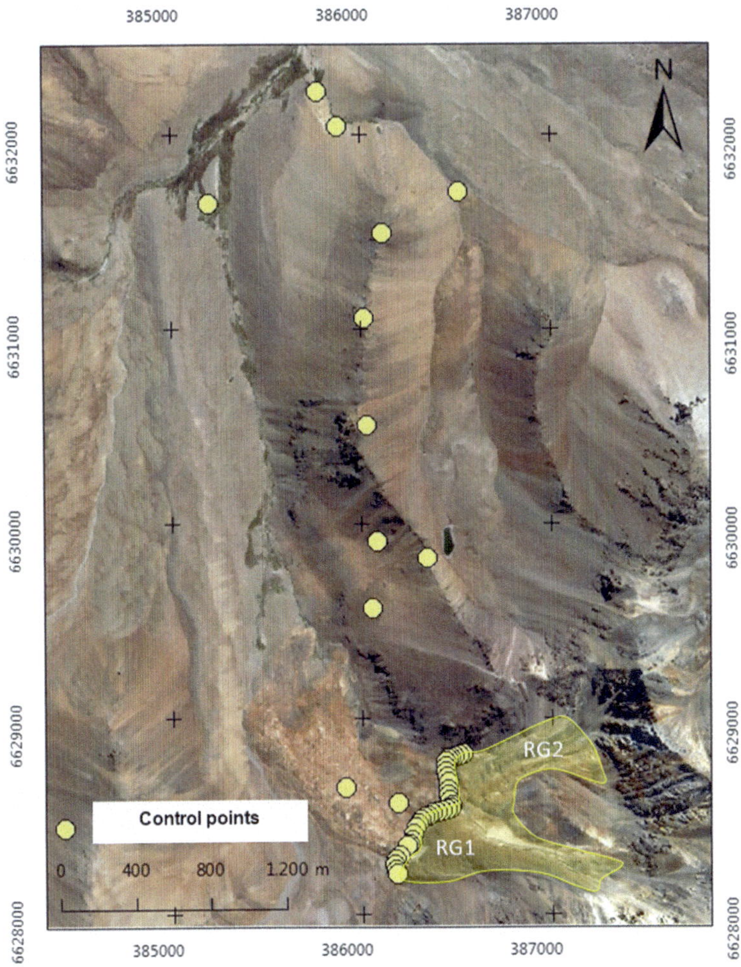

Fig. 3.5 Control points in the terrain (GCPs), taken in outcrops of the Quebrada Caballos and at the foot of the rock glaciers 1 and 2. In yellow, the contours of the rock glacier 1 and rock glacier 2. Base map: Worldview ArcGIS, February 2015. Coordinate system UTM 19 S, Datum WGS 84

The existence of a large debris accumulation at its base indicates that they are being affected presently by cryoclastism, by mass-movement processes and fluvial action, whereas the good development of their morphology (U-shaped valleys, polished bedrock, well-preserved moraines) indicates a prolonged evolution under glacial-cryogenic conditions. Most of them are in contact with rock glaciers at their base.

Likewise, the basal moraines are the most abundant landforms in the study zone. They cover the base of the glacial valleys of Quebrada Caballos, Toro Muerto and Vallecillo. This type of moraine is the best preserved in the area, frequently presenting

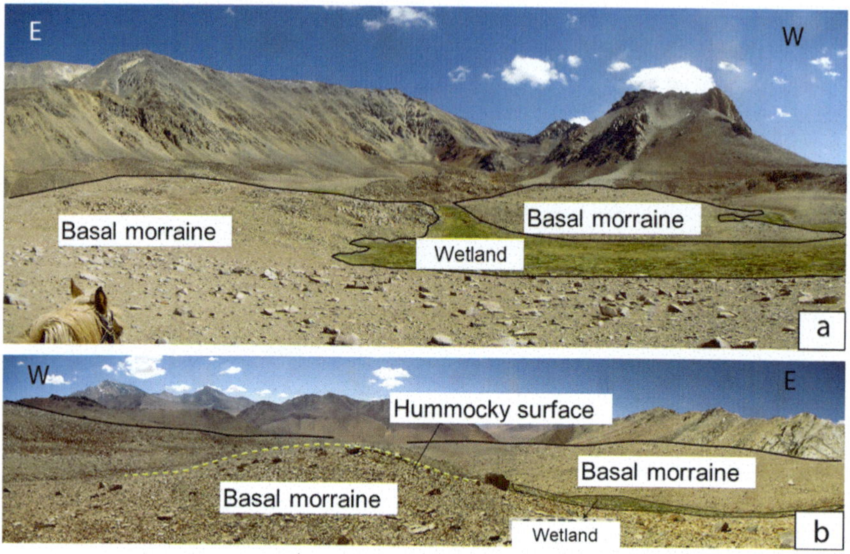

Fig. 3.6 Basal moraine belonging to the joint of the Toro Muerto and Vallecillo Valleys. **a** Basal moraine cut down by creeks, in whose sides wetlands have been developed. **b** View towards the north from the surface, where it may be observed the surface, where the hilly landscape of the deposit can be appreciated

work by fluvial processes (Fig. 3.6a). The surface is characterized by a hilly topography of "hummock" type (Fig. 3.6b), and the deposit, of similar shape to the other types of moraines, is greyish, poorly sorted and matrix-supported, with blocks of more than 1 m in diameter faceted and polished in the higher sectors.

An orangish deposit is located at the foot of the rock glaciers 1 and 2 of the Quebrada Caballos, at 3,978 m a.s.l. with an area of 0,68 km^2 and 1,229 km length (Fig. 3.7a). The shape of this deposit was probably generated by a rock avalanche coming from the SE. Due to its position in the valley, it is possible to infer that it corresponds to an event older than the formation of the two rock glaciers (Fig. 3.7b) which could occur due to a catastrophic rock flow of the pre-existing, weathered rock slopes. However, it is necessary to point out that due to its morphological and depositional complexity, it is difficult to characterize it as a rock avalanche.

One of the important results of this work was the updating of the geomorphological map of the study area, based on all the observations presented in this chapter, considering geomorphologic, cryogenic and glacigenic features in the High Andean Cordillera, which is shown in Fig. 3.8.

In addition, as a result of the geomorphological mapping and in order to complement the public inventory of glaciers of the DGA (2015), 11 rock glaciers have been identified and classified as active or inactive (Fig. 3.9). They have a total surface of 1.27 km^2, concentrated above 4,000 m a.s.l. in the bottom valleys of Caballo and Vallecillo with abrupt rock scarps. Different rock glacier types are recognized taking into consideration such parameters as morphology and position.

Fig. 3.7 **a** View in plant of the mass-movement deposit identified in the bottom of the Quebrada Caballos. **b** View towards the E of the scarp where it could have generated the movement, noticing also the two rock glaciers of the valley. **c** Hilly, hummocky surface of the avalanche deposit. **d** Front of the deposit, with greyish boulders exposed

The surface of these bodies is covered by coarse materials, and they present a surface with crests and curved transverse depressions. One of the more remarkable characteristics, especially observed in the rock glaciers 1 and 2, would correspond to surficial runoff coming from the foot of both landforms, including also in situ evidence of runoff/water flow to the inner portion of the body off the rock glacier 1.

In addition, extensive slopes characterized by the presence of lobes and wrinkles are predominantly located in higher zones (above 4,100 m a.s.l.). These slopes are thus observed because they are directly related to the soil grain size, which is generally of affine grained type. This happens because, if there is enough snow that can subsequently melt (and recrystallize to deepen as ice in winter) this surface becomes inflated in a perpendicular manner to the ice, whereas when the soil is seasonally unfrozen, the surface is almost parallel to the force of gravity, that is, it occurs vertically, showing such deformation (French 2007). The lobes are the resulting landforms, being along the length and width of the slope.

Cryoweathering is a process that occurs in the rock outcrops, being more evident in the Vallecillo Valley (Fig. 3.10). It corresponds to the mechanical disintegration of the soil or rock, as a consequence of the processes of freezing and melting, ice pressure and hydration (Trombotto et al. 2014).

The selection of loose materials constitutes one of the most notorious evidences of the action of water freezing and melting of the ice in cryogenic environments.

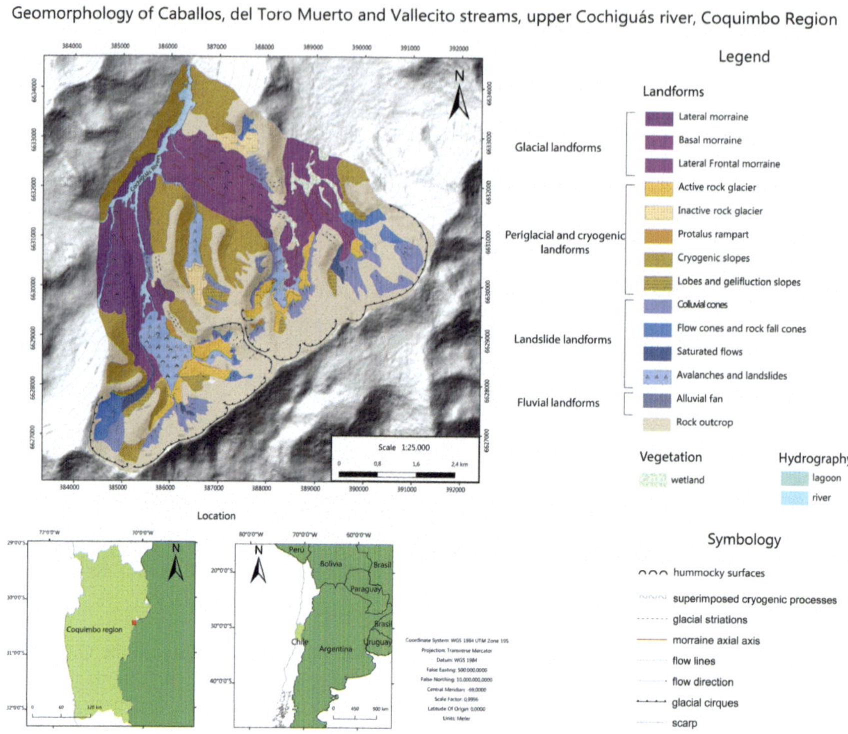

Fig. 3.8 Updated geomorphological map of the study zone considering periglacial and cryogenic landforms

Evidence of landforms affected by these processes correspond to the southwestern lateral moraine located south of the Quebrada Caballos (MLW) and the basal moraine of the Quebrada del Toro Muerto.

According to the presence of ice, in the upper portion of the rock glacier 2, at almost 4,270 m a.s.l., a groove containing refrozen ice in its inner core has been dug up to 50 cm from the surface. The debris cover contained is coarse, the size of the blocks oscillates between 20 and 30 cm and even bigger ones, reaching sizes of up to 70 cm in some cases. The ice is massive and translucid, and it contains inside, clasts of smaller size, frozen and united by the ice. It is found in the sedimentary matrix of the debris cover cementing the clasts of smaller size, thus being classified as cement ice or matrix ice.

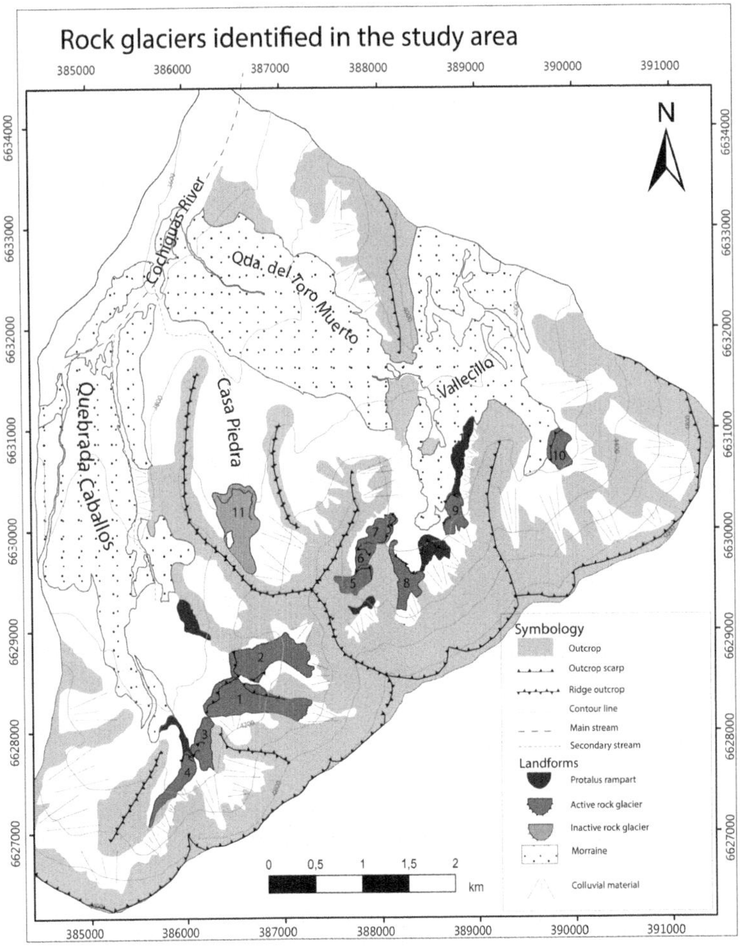

Fig. 3.9 Updated rock glacier inventory for the study area, and associated landforms and geomorphological processes

3.6 Rock Glaciers Characteristics

The delimitation of the basal and surface contours of rock glacier 1 and 2 was carried out from GPS points taken on site and by remote observations due to the complexity of the terrain, which made it impossible to take data in situ. The result of the contours obtained is shown in Fig. 3.11.

Thus, it is possible to point out that the rock glaciers have a multilingual surface shape, and their upper part is connected directly from the base of the rock walls (cirques) at the bottom (in the case of rock glacier 2) or very close to them (rock glacier 1).

Fig. 3.10 Basal moraine of the Quebrada del Toro Muerto. Towards the left, satellite image of the deposit, where the surfaces in black are delimiting the surfaces with special features. To the right, photographs of the deposit in different angles, evidencing the sorting recurrence due to freezing in depressions of the deposit, which concentrate most part of the oxidized, orange clasts

On the other hand, both are approximately 1 km long and have an area of 0.22 km^2 (rock glacier 2) and 0.26 km^2 (rock glacier 1). They are oriented E-W (rock glacier 1) and SW (rock glacier 2) and have a similar frontal angle, between 35° and 37°. However, there are parts of the front that reach values greater than 40°, especially in rock glacier 2.

The rock glacier 1, located under the southernmost rock slope at the bottom of Caballos Creek, shows typical compression features such as transverse ridges along the maximum length of the body, as well as grooves or depressions towards the end of the body. The latter are the most visible in satellite images and are found between cords and ridges, also transversal (Fig. 3.12). These forms of compression are more recurrent in the most distal part of the glacier, in the area with the lowest surface slope. There are also longitudinal depressions and ridges, but these are not as recurrent as the transverse type, and are located in the central-high sector of the body, at the lateral ends of the body (Fig. 3.12).

On the other hand, the rock glacier 2 is located immediately under rock slopes with southern exposure and SW exposure, giving this landform a preferential orientation

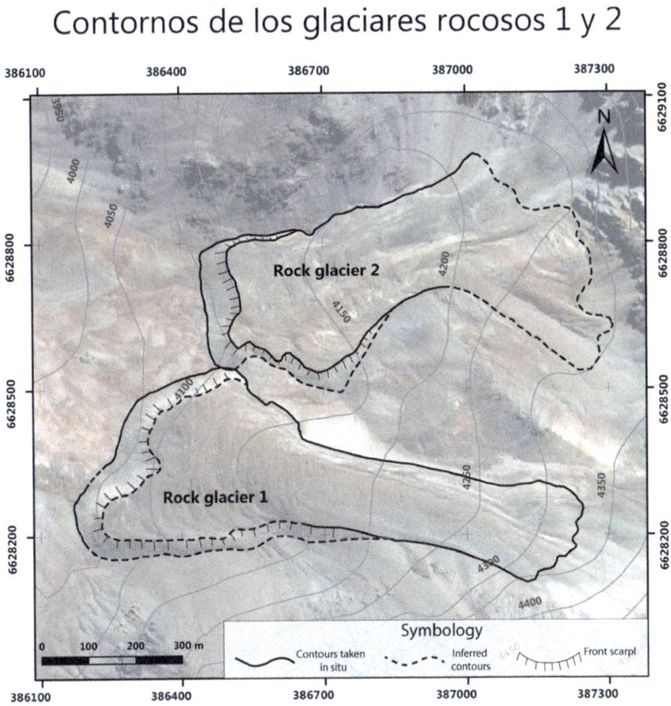

Fig. 3.11 Contours of rock glaciers 1 and 2 delimited by GPS points taken in the field and in remote observations. Dashed line represents inferred areas due to terrain complexity. Base map: Worldview ArcGIS February 2015. UTM 19 S coordinate system, WGS 84 datum

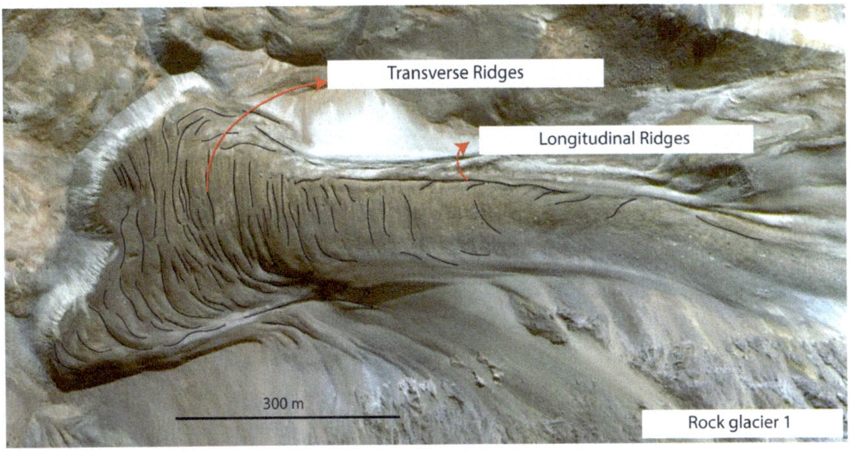

Fig. 3.12 Plan view of rock glacier 1, where the ridges visible in the satellite image, which are concentrated in the lower limit of the body, have been marked with a black line. Base map: May 7, 2008 image from the Google Earth server

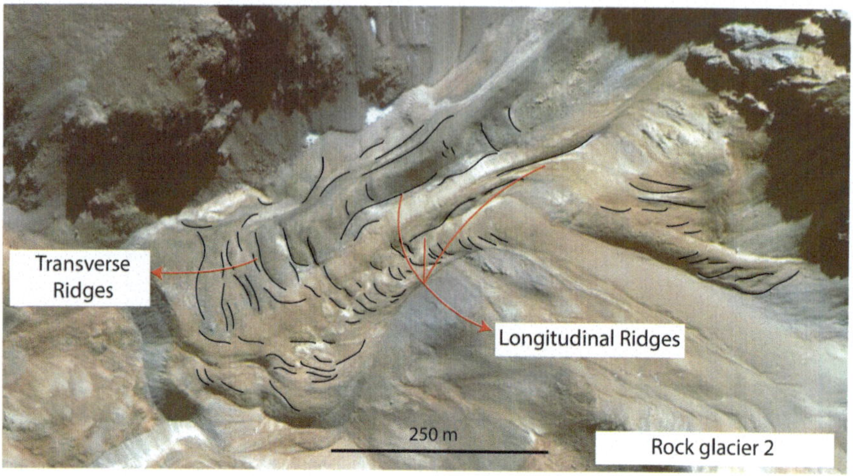

Fig. 3.13 Satellite image of rock glacier 2 showing the main longitudinal and transverse ridges identified. Base map: May 7, 2008 image from Google Earth server

towards the SW. Morphologically, it is different from rock glacier 1, because its surface is characterized by pronounced and deep depressions both transversely and longitudinally, such as those shown in Fig. 3.13.

However, contrary to what was observed in rock glacier 1, the grooves or depressions resulting from compressional processes are not so numerous or so close to each other, and are located mainly in the lower zone of the glacier. There are abundant longitudinal depressions located mainly in the centre of the body, which also mark an evident change in the surface coloration of the detrital cover. The surface shape is not as elongated as the neighbouring glacier, and its source of nourishment apparently comes directly from rock outcrops located near its upper boundary.

Important characteristics of this glacier include the presence of cryogenic features such as cryometeorization, circular depressions, an irregular cover characterized by deep longitudinal grooves up to 10 m deep, the presence of internal ice under the detrital cover in the upper zone of the rock glacier, snow forms such as penitentes and terraced snow covered by detritus.

The presence of snow was found in the upper portions of the rock glacier 2 in the shape of penitents, with a height between 10 and 50 cm. They have also been observed in nivation niches above the rock glacier 2, covering an area of small dimensions. Additionally, at the upper boundary of the glacier, surface snow has been found in the shape of terraces, which are presently covered by a thin layer of fine-grained debris, of approximately 5 cm in thickness. On these snow terraces, it is possible to observe debris of larger sizes, distributed in a chaotic manner. These debris of larger dimensions are also under snow benches. The height of each snowbank is approximately 40 cm and the snow that composes them is soft and dirty. This buried snow would correspond to multiannual snow, which could be related with the surficial ice, as noted before.

Concerning the historical variation and estimated volume of the rock glaciers of Quebrada Caballos, using the analysis of the digital elevation model (1955–2018) it was found that the average altimetric variation for the rock glacier 1 is −2.56 m with a standard deviation of 6.34 m, whereas for the rock glacier 2, the average value of the altimetric variation corresponds to −10.26 m, with a standard deviation of 5.60 m. Therefore, the average of the altimetric difference for this period for both glaciers is −6,41 m, which represents an average diminution of 10.17 cm per year. In this way, the total variation for the 1955–2018 period corresponds to −6,41 m, that is, −10 cm per year.

Assuming a highly conservative condition of total saturation of interstitial spaces in the ice, the annual variation is of approximately 27,000 m^3. This is equivalent to a fluvial runoff of 0.002 m^3/s considering four melting months (summer season), which is an insignificant value in terms of error ranges of discharge measuring methodologies.

3.7 Discussion

This high mountain landscape in the eastern slope of the Central Andes is the result of exogenic agents during the Quaternary, both of glacial and periglacial type. Moreover, the posterior modelling associated with the interaction of the periglacial system with gravitational processes, cryoclastism, seismicity, fluvial, debris transportation and of chaotic torrential character (such as mixed cones and saturated flows) rules the local morphodynamics at present.

The magnitude of the erosional processes associated with the ice in the past has been influenced by the structural characteristics of the zone, the relief orientation and by the rock competence due to its degree of fracturing. According to Paskoff (1970), those are key factors in the modelling of the relief. This would largely explain the magnitude of the erosional landforms as well as those of accumulation landforms such as moraines and talus cones distinguished both along the entire Cordillera del Elqui (Paskoff 1970).

The moraines of the study zone, especially those of the lateral type, are noticeable by their symmetry, similarities in the composition of their deposits and a very well-preserved morphology. This "perfection" in the symmetry of the moraines has been attributed by Paskoff (1970) to a pre-glacial condition, where the glaciers would have flown along valleys which were very wide already, since from a beginning due to the uplifting of the mountain chains in the third epoch of the Andean Cycle, in a manner that they were retouched by the action of the ice and later in-filled with an appropriate sedimentary load/discharge.

The glacial valleys that dominate the study area are mostly of short dimensions and quite wide at its base.

In relation to the landforms generated by mass-movement processes, the abundance of mixed talus cones has been originated from the fragmentation of the rocks due to cryoclastism, a process that increases with altitude and that generates abundant landforms of this type along the mountain systems (Paskoff 1970). In a general way, the slopes free of ice provide abundant debris for the development of cryogenic rock glaciers, being these of the talus or valley bottom types (Brenning et al. 2007).

The geomorphological map obtained is mostly coincident with those shown by Paskoff (1967). However, there exist several differences that keep relationships with the scale at which the mapping has been done (1:25.000 vs. 1:250.000), as well as the omission of other morphologies due to their relationship with other processes. In this sense, Paskoff (1967) indicated that the area is dominated by landforms and deposits of glacial or nival origin, as these processes are the principal modelling agents of the landscape. Notwithstanding, the presence of other morphologies linked to gravitational, fluvial or periglacial is not indicated but they are widely recognized in the study area.

The glacial evidence presented indicates that colder climates occurred in the past, allowing the development of valley glaciers whose tongues would have reached up to the confluence of the Río Claro with the Río Cochiguás, partly evidenced by the basal moraines located in Toro Muerto, Vallecito and Quebrada Caballos valleys (Figs. 3.6 and 3.10). Additionally, the location of the rock glaciers of the Quebrada Caballos and some of the Quebrada Vallecillo allows their classification as valley bottom, rock glaciers, whose origin is associated with glacial processes (Humlum 1982). This permits to suggest that these landforms are morphogenetically related to glacial processes in its origin, and to periglacial processes in their evolution, constituting a continuous geomorphic system (Soto et al. 2002). In this manner, it is possible to remark that the main rock glaciers studied (rock glaciers 1 and 2) are glacigenic rock glaciers, whose origin would be associated with ice bodies existing in the past in the semiarid Andes (Riquelme et al. 2011). The geomorphological sequence of the construction of the landscape in the Quebrada Caballos may be proposed as follows:

1. A first step related to a glacial advance responsible for the deposition of the lateral moraines of the valley (MLW and MLE) and of the latero-frontal moraine. These moraines are located at high elevations and are morphologically well preserved that would allow to assign them to the Last Glacial Maximum (32 ka approximately, Zech et al. 2007) or to an episode of glacial advance reported for the semiarid Andes between 17 and 12 ka B.P. (Zech et al. 2006, 2007, 2008).
2. A second phase of glacier retreat due to more arid conditions during the Holocene (Ammann et al. 2001; Grosjean et al. 1998; Riquelme et al. 2011) gave path to periglacial conditions, increase in the instability of the slopes due to deglaciation at the same time of an increase in the debris production and valley in-filling forming dejection cones and debris cones, due to the transition towards a more arid, warm environment, conditioned to freezing process and seasonal melting.
3. A last phase of complete transition towards a full periglacial environment with the predominance of cryogenic features and landforms and rock glaciers, both glaciogenic and cryogenic. Under these conditions, the rock glaciers accumulated snow meltwater that refreezes forming ice in the superficial ice/rock matrix, afterwards liberated, at least partially, during the summer (Corte 1988).

The identification of 11 rock glaciers with varied morphological differences among them and different placement sites allows to infer the existence of landforms of different origin, coeval with the same morphoclimatic domain, obeying singular topoclimatic and geomorphological conditions that have favoured their formation and development.

This number of active and inactive rock glaciers differs with the number indicated by the National Glacier Inventory (DGA 2015) (Fig. 3.14). In this case, only three active rock glaciers were found in the studied valleys, corresponding to the rock glaciers 1, 2, and 4 of the inventory performed in this present study (Fig. 3.9). In the terrain, it was possible to verify that the remaining landforms numbered as 3, 5, 6, 7, 8, 9 and 10 (Fig. 3.9) actually correspond to rock glaciers, making it very probable that the methods used in the former inventory from DGA would not be the most appropriate ones, for instance, the use of old satellite imagery of low resolution makes impossible a good definition of the landforms.

It was also observed that in both glaciers there is an important supply of debris from the lateral slopes of both rock glaciers, which is especially evident in the upper zone of rock glacier 2, due to the presence of talus cones, debris flows and rockfalls. In this sense, these glaciers play an important role in the system of debris transportation (Brenning 2005), since the presence of these landforms not only stabilizes the debris material in itself, but also constitutes a base level for fast mass-movement processes as avalanches, rock falls and even small landslides (Brenning 2005).

Concerning their thinning, rock glacier 2 would be characterized by a gradual but continuous reduction of the inner ice, where the amount of internal debris would become gradually proportional to the ice quantity. It will be completely covered by debris and ice would not be observable at the surface (only buried), something that was corroborated in the terrain and typically holding values of 25–45% of ice. Among the additional characteristics that are coincident with those observed in the rock glacier 2, there are pronounced crests and grooves both transversal and longitudinal, in addition to the longitudinal groove. This last one would indicate the reduction of the ice core in this rock glacier. According to Janke et al. (2015), the consequence due to an atmospheric warming in this type of glaciers is the continuous reduction of the ice under the surface, which would make slow the movement and would reduce its capacity as a water reserve. In a segregated or interstitial manner, the front appears more or less pronounced and elongated, whereas the transversal chains are observed to be more of a lineal type.

To be able to establish the quantitative supply of the rock glaciers and the actual ice content in their inner portion, it is needed to make later studies that would include, for instance, coring combined with geophysical and geochemical studies, so as to obtain real data related to the inner structure, water provenance and full water supply.

Although the model performed to evaluate the rock glacier variations along time is not the most precise, it indicates tendencies that were already estimated from time ago in those studies related to mountain permafrost (Mollaret et al. 2019). Therefore, to improve even more the precision and results, it is needed to perform more studies in the semiarid zone of Chile, following the vertical variations using for that purpose

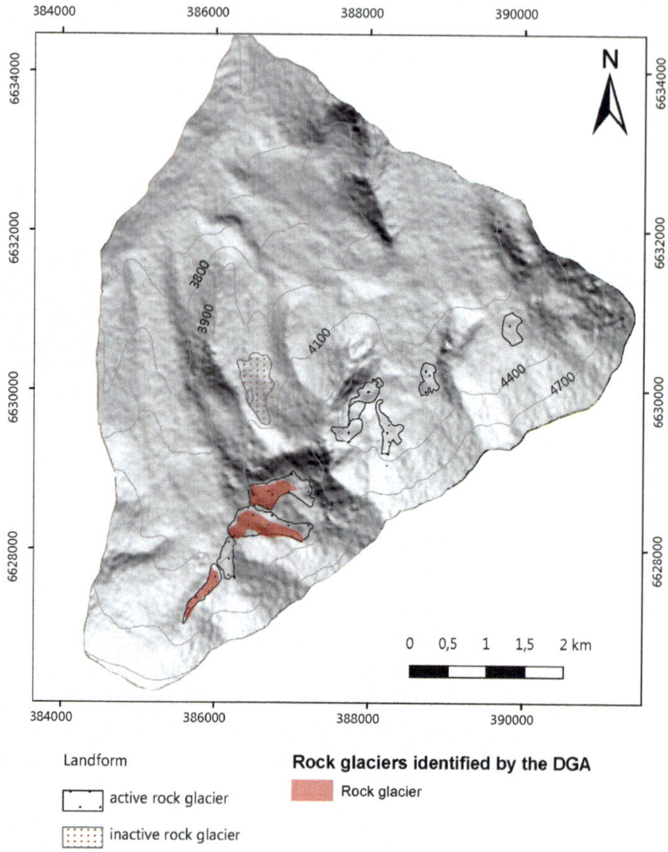

Fig. 3.14 Rock glaciers (active and inactive) identified in this study. The glaciers also identified in the National Glacier Inventory (Inventario Nacional de Glaciares, DGA 2015) are in red. Base Map: hillshade developed from DEM SRTM 2000, contours of IGM. System of coordinates UTM 19 S, datum WGS 84

more sophisticated methodologies, as drones, high-resolution imagery, techniques of Structure from Motion (SfM), etc.

The values obtained indicate that there is a mean thinning of the ice of 10 cm per year in the period between 1955 and 2018. However, the loss rates per year vary for each glacier. In fact, rock glacier 1 has experienced a thinning of 4 cm per year, whereas rock glacier 4 has experienced a thinning of 16 cm per year. These results are very different in between them, and they differ in different periods, especially in the period 1999–2018, when it has been observed a mean increase in thickness in the 1 m vertical of rock glacier 1, and a drastic thinning of the mean elevation of rock glacier 2, with a value close to—22 cm per year. These differences in the mentioned period may be explained mainly by: (1) a larger debris cover in rock glacier 1 which allows for a better protection (Lambrecht et al. 2011), (2) by methodological mistakes in the photogrammetric process or 3) because the period comprised (1999–2018) is not enough to find reliable results.

According to the literature, the active rock glaciers are preferably located in zones where the mean annual temperature is lower than −1 or −2 °C and where the precipitation does not exceed 2,500 mm/year (Haeberli 1985; Barsch 1996). In the case of the studied valleys, the origin area of all rock glaciers is located at a mean elevation of 4,400 m a.s.l., with a mean annual temperature equal or lower than −2 °C, coincidently with that stated above, the frontal zone is approximately located at 4,100 m a.s.l., in zones where the mean annual temperature is slightly above 0 °C. Consequently, the distribution of active rock glaciers over this elevation occurs with a mean annual temperature of at least 0 °C. At this latitude, these morphologies are common, even in the cirques exposed towards the NW or NE, with a high input of solar radiation (Brenning 2005).

From these observations, it is possible to state that the active and inactive landforms between 3,900 and 4,400 m a.s.l. in this high mountain sector may reflect past thermal conditions, due to the large time response scale of the rock glaciers and of the permafrost in general (in the order of decades and centuries, after Haeberli (1990). This would suggest an increase of at least 1 °C due to climatic changes, which would not impact in great manner on the rock glaciers situated under the −2 °C isotherm, since they would be kept under cold conditions. However, the rock glaciers located in isotherms between 0 and −1 °C (three of them and the front of the rock glaciers 1 and 2) could develop a greater sensitivity to temperature increase and due to this warming could cause more melting (Azócar 2013).

It is especially important the presence of fresh snow terraces covered by debris, the penitents of ice and snow identified in the upper part of rock glacier 2. This snow is probably coming from avalanches in the higher portions, and it would allow to explain, partially, the incorporation of snow inside this part of the glacier due to the melting dynamics and inner refreezing, also providing evidence of the fast debris production on the slopes that occur at the heads of the valleys and which determines the dynamics of these landforms. This indicates the importance of snow precipitation that would favour the supply of summer discharge in the zone, forming part of the ice contained in the interior of the debris cover, which freezes and refreezes seasonally.

Besides, in the surroundings of both rock glaciers, and particularly in rock glacier 1, continuous or discontinuous snow was not observed, but the presence of inner runoff in the glacier was seen, what allows to infer that due to a greater incident solar radiation with respect to the rest of the study zone, there is a greater rate of snow melting that runs along the inner body of the rock glacier.

Taking into consideration the studied period, the thinning rates observed in the rock glaciers 1 and 2 would be responding, partly, to historical tendencies of atmospheric warming and diminution of precipitation, especially stressed during the last decades due to the ENSO phases and the aridification of the Central Andes (Santibañez 1997; Le Quesne et al. 2006; Vuille and Milana 2007).

However, if the precipitation deficit close to 30% since the year 2010 is considered related to drought events of diverse severity, intensity and duration, a phenomenon that has been named as the "megadrought of Central Chile" by Aldunce and González

(2009) and Garreaud et al. (2017). According to it, a smaller quantity of snow would have influence in small recharge and thermal isolation for the underlying ice, which, together with a regular increment of the temperatures, would affect at least the more superficial layer of intrinsic ice of the glacier. This cause/consequence relationship is applicable to precipitation, therefore, to the negative historical variation of both rock glaciers in the studied period, understanding that these climatic tendencies have been observed since past decades.

Clearly, the rock glaciers respond in a different manner depending upon the thermal régime, the microclimate conditions, the debris supply, the topography and geological parameters (Arenson and Jakob 2010). In spite of the fact that the Central Andes seem to offer optimal conditions for the development of rock glaciers, it is also susceptible to atmospheric warming so a pronounced and generalized change is expected in the rock glaciers existing in this sector. The morphological changes of these glaciers in a warmer climate are still unknown, although more of them would become relics once that their boundaries stay below the 0 °C isotherm (Bodin et al. 2012; Iribarren Anacona et al. 2014).

Finally, it is estimated that in the near future, the atmospheric warming would elevate the 0 °C isotherm, around 500 m above the present position in the worst scenario, even more than during the Holocene. Therefore, it is expected that in the Central Andes, close to the 95% of the active ice-content landforms would have its terminal position below the 0 °C isotherm around 2070 (Drewes et al. 2018).

3.8 Conclusions

The geomorphological characteristics of the high basin of the Río Cochiguás clearly show how the different morphogenetic agents have had influence in different time scales on the present configuration of the landscape. This is reflected in the existence of landforms directly attributed to the glacial action, as cirques and glacial valleys, that presently show in their surface some of the most common features of the erosion systems of periglacial environments.

It is also recognized the importance to have available a precise geomorphological map at the time to suggest the uses of the territory in the mountain areas. Particularly, because it was determined that in the same morphoclimatic domain or altitudes, there may exist a great variety of processes depending on the mixing of multiple factors.

The previous information cited here, as well as the relationship between morphological, geological, topoclimatic and regional characteristics in the evolution and present historical variation of the rock glaciers, are theoretical self-supported. Nevertheless, it is needed the exploration and study of these landforms in mountain areas of large dimensions and with great geological variations to validate these results at the scale of a semiarid region. In this sense, there is still much to do about the real water implication of these cryoforms and the impact that the climatic change has on them, as well as about their dynamics and possible mechanisms of feeding/melting, what has a clear relationship with the interaction between solid precipitation, the resulting warming and the role of the different thickness of the debris covers, as they act as a thermal transmitter or a thermal separation.

3.9 Limitations and Updatings

This research was performed in 2018. Until today (2022), the International Permafrost Association (IPA) has developed new lines of research and technical handbooks for the development of inventories of rock glaciers under new morphological standards that may differ from what has been presented in this chapter.

Acknowledgements Special thanks to the Servicio Nacional de Geología y Minería of Chile (SERNAGEOMIN) for having financed this research in the updating of the Elqui–Paso de Agua Negra geological sheet. The academics Gino Casassa, Guillermo Azócar and Gabriel Easton were part of the bachelor thesis committee in which this topic was developed.

References

Aguilar G, Riquelme R, Martinod J, Darrozes J, Maire E (2011) Variability in erosion rates related to the state of landscape transience in the semi-arid Chilean Andes. Earth Surf Proc Land 36:1736–1748

Aguilar G, Riquelme R, Martinod J, Darrozes J (2013) Role of climate and tectonics in the geomorphologic evolution of the Semiarid Chilean Andes between 27 and 328S. Andean Geology 40:79–101

Aldunce P, González M (2009) Desastres asociados al clima en la agricultura y medio rural en Chile. FIA, Chile, p 117p

Ammann C, Jenny B, Kammer K, Messerli B (2001) Late quaternary glacier response to humidity changes in the arid Andes of Chile (18–29°S). Palaeogeogr Palaeoclimatol Palaeoecol 172:313–326

Arenson LU, Jakob M (2010) The significance of rock glaciers in the dry Andes—A discussion of

Azócar G (2013) Modeling of Permafrost Distribution in the Semi-arid Chilean Andes. M.Sc Thesis, University of Waterloo. Waterloo, Ontario, Canada

Ballantyne CK (2002) Paraglacial geomorphology. Quat Sci Rev 21:1935–2017

Barazangi M, Isacks BL (1976) Spatial distribution of earthquakes and subduction of the Nazca plate beneath South America. Geology 4(11):686–692

Barsch D (1978) Active rock glaciers as indicators for discontinuous alpine permafrost. An example from the Swiss Alps. Third International Conference on Permafrost, vol 1. Ottawa, pp 349–353

Barsch D (1996) Rock glaciers. Indicators for the present and former geoecology in high mountain environments. Springer Verlag, Heidelberg

Bissig T (2001) Metallogenesis of the Miocene El Indio-Pascua gold-silvercopper belt, Chile/ Argentina: geodynamic, geomorphological, and petrochemical controls on epithermal mineralisation: Ph. D. thesis, Kingston, Canada, Queen's University, 531 p

Brenning A (2005) Geomorphological, hydrological, and climatic significance of rock glaciers in the Andes of Central Chile (33–35 S). Permafrost Periglac Process 16:231–240

Brenning A, Grasser M, Friend D (2007) Statistical estimation and generalized additive modeling of rock glacier distribution in the San Juan Mountains, Colorado, USA. J Geophys Res Earth Surf 112:F02S15

Bodin X, Krysiecki JM, Anocona PI (2012) Recent collapse of rock glaciers: two study cases in the Alps and in the Andes. In: 12th congress, protection of living spaces from natural hazards, Interpraevent 2012, vol 1, p 409
Charrier R, Baeza O (2002) Evidence for Cenozoic extensional basin development and tectonic inversion south of the flat-slab segment, southern Central Andes, Chile (33–36S.L.). J S Am Earth Sci 15:117–139
Corte AE (1987) Central Andes rock glaciers: applied aspects. In: Giardino JR et al (eds) Rock glaciers. Allen and Unwin, London, pp 289–304
Corte AE (1988) Geocryology of the Central Andes and rock glaciers. In: Fifth international conference on permafrost, proceedings, vol 1. Tapir Publishers, Trondheim, Norway, pp 718–723
DGA-Dirección General de Aguas (2015) Glaciares Nacionales al Año 2015. Versión en línea. http://www.dga.cl/estudiospublicaciones/mapoteca/Paginas/default.aspx#cinco
Drewes J, Moreiras S, Korup O (2018) Permafrost activity and atmospheric warming in the Argentinian Andes. Geomorphology 323:13–24
Eriksen CC, Blumenthal MB, Hayes SP, Ripa P (1983) Wind-generated equatorial Kelvin waves observed across the Pacific Ocean. J Phys Oceanogr 13(9):1622–1640
Errázuriz A, Cereceda P, González M, Henríquez M, Rioseco R (1998) Manual de Geografía de Chile. Ed. Andrés Bello. Santiago, 433p
Escobar F, Aceituno P (1998) Influencia del fenónemo ENSO sobre la precipitación nival en el sector Andino de Chile Central durante el invierno. Institut Français D'études Andines 27(3):753–759
Fiebig-Wittmaack M, Astudillo O, Wheaton E, Wittrock et al (2012) Climate trend and impact of climate change on agriculture in an arid Andean valley. Climate Change 111(3–4):819–833.
Folguera A, Ramos VA, Melnick D (2002) Partición de la deformación en la zona del arco volcánico de los Andes neuquinos (36–39 S) en los últimos 30 millones de años. Revista Geológica De Chile 29(2):227–240
French HM (2007) The periglacial environment. Wiley, West Sussex, England
Gascoin S, Kinnard C, Ponce R, Lhermitte S, MacDonell S, Rabatel A (2010) Glacier contribution to streamflow in two headwaters of the Huasco River, Dry Andes of Chile. Cryosphere Discuss 4:2373–2413
Garreaud R, Alvarez-Garreton C, Barichivich J, Zambrano-Bigiarini M (2017) The 2010–2015 megadrought in central Chile: impacts on regional hydroclimate and vegetation. Hydrol Earth Syst Sci 21(12):6307–6327
Garreaud R, Rutllant J (1997) Precipitación estival en los Andes de Chile central: aspectos climatológicos. Atmósfera 10(4):191–211
Giardino JR, Vitek JD (1988) The significance of rock glaciers in the glacial–periglacial landscape continuum. J Quat Sci 31:97–103
Ginot P, Kull C, Schotterer U, Schwikowski M, Gaeggeler HW (2006) Glacier mass balance reconstruction by sublimation induced enrichment of chemical species on Cerro Tapado (Chilean Andes). Clim Past 2(21–30):2006
Grosjean M, Geyh MA, Messerli B, Schreier H, Veit H (1998) A late-Holocene (<2600 BP) glacial advance in the south-central Andes (29 S), northern Chile. Holocene 8(4):473–479
Gutscher MA, Maury R, Eissen JP, Bourdon E (2000) Can slab melting be caused by flat subduction? Geology 28(6):535–538
Haeberli W (1985) Creep of mountain permafrost: internal structure and flow of alpine rock glaciers. Mitteilungen Der Versuchsanstalt Für Wasserbau, Hydrologie Und Glaziologie 77:142
Haeberli W (1990) Pilot analysis of permafrost cores from the active rock glacier Murtel I, Piz Corvatsch, Eastern Swiss Alps—A Workshop Report. Hydrologie und Glaziologie, ETII, vol 9, Zürich, 38 p
Haeberli W, Hallet B, Arenson L, Elconin R, Vonder Mühll D (2006) Permafrost creep and rock glacier dynamics. Permafrost Periglac Process 17(3):189–214
Humlum O (1982) Rock glacier types on Disko, central west, Greenland. Danish J Geogr 82:59–65
Humlum O (1988) Rock glacier appearance level and rock glacier initiation line altitude: methodological approach to the study of rock glaciers. Arct Alpine Res 20:160–178

Iribarren Anacona P, Mackintosh A, Norton KP (2014) Hazardous processes and events from glacier and permafrost areas: lessons from the Chilean and Argentinean Andes. Earth Surface Process Landforms 21:2–21

Ikeda A, Matsuoka N (2006) Pebbly versus bouldery rock glaciers: morphology, structure and processes. Geomorphology 73(3–4):279–296

Janke JR, Bellisario AC, Ferrando FA (2015) Classification of debris-covered glaciers and rock glaciers in the Andes of central Chile. Geomorphology 241:98–121

Janke JR (2013) Using airborne LiDAR and USGS DEM data for assessing rock glaciers and glaciers. Geomorphology 195:118–130

Jara P, Charrier R (2014) Nuevos antecedentes estratigráficos y geocronológicos para el Meso-Cenozoico de la Cordillera Principal de Chile entre 32° y 32° 30′S: Implicancias estructurales y paleogeográficas. Andean Geol 41

Johnson PG (1984) Rock glacier formation by high-magnitude low-frequency slope processes in the Southwest Yukon. Ann Assoc Am Geogr 74:408–419

Jordan TE, Isacks BL, Allmendinger RW, Brewer JA et al (1983) Andean tectonics related to geometry of subducted Nazca plate. Geol Soc Am Bull 94(3):341–361

Lambrecht A, Mayer C, Hagg W, Popovnin V, Svanadze D (2011) A comparison of glacier melts on debris-covered glaciers in the northern and southern Caucasus. Cryosphere 5:525–538

Le Quesne C, Stahle DW, Cleaveland MK, Therrell MD, Barichivich J (2006) Ancient *Austrocedrus* tree-ring chronologies used to reconstruct central Chile precipitation variability from A.D. 1200–2000. J Climate 19(22):5731–5744

Maksaev V, Moscoso R, Mpodozis C, Nasi C (1984) Las unidades volcánicas y plutónicas del Cenozoico superior entre la Alta Cordillera del Norte Chico (298–318S), geología, alteración hidrotermal y mineralización. Revista Geológica De Chile 21:11–51

Mollaret C, Hilbich C, Pellet C, Flores-Orozco A, Delaloye R, Hauck C (2019) Mountain permafrost degradation documented through a network of permanent electrical resistivity tomography sites. Cryosphere 13:2557–2578. https://doi.org/10.5194/tc-13-2557-2019

Monnier S, Kinnard C, Surazakov A, Bossy W (2014) Geomorphology, internal structure, and successive development of a glacier foreland in the semiarid Chilean Andes (Cerro Tapado, upper). Geomorphology 207:126–140

Mpodozis C, Cornejo P (1988) Hoja Pisco Elqui. IV Región de Coquimbo. Servicio Nacional de Geología y Minería, Carta Geológica de Chile, No. 68, Santiago

Mpodozis C, Brockway H, Marquardt C, Perello J (2009) Geocronología/Pb y tectónica de la región de Los Pelambres-Cerro Mercedario: implicancias para la evolución cenozoica de Los Andes del centro de Chile y Argentina. In: XII Congreso Geológico Chileno, 12, Santiago, Chile, Departamento de Geología de la Universidad de Chile

Murillo I, Velásquez R, Creixell C (2017) Geología de las cartas Guanta – Los Cuartitos y Paso de Vacas Heladas, Regiones de Atacama y Coquimbo. Servicio Nacional de Geología y Minería, Carta Geológica de Chile, Serie Geología Básica Nos. 192–193: 96 p., 1 mapa escala 1:100.000, 1 CD con anexos. Santiago

Nasi C, Moscoso R, Maksaev V (1990) Hoja Guanta, Regiones de Atacama y Coquimbo, Sernageomin, Santiago, Chile

Ota Y, Miyauchi T, Paskoff R, Koba M (1995) Plio-quaternary terraces and their deformation along the Altos de Talinay, North-Central Chile. Revista Geológica De Chile 22:89–102

Pardo M, Vera E, Monfret T, Yánez G, Eisenberg A (2006) Sismicidad cortical superficial bajo Santiago: implicaciones en la tectónica andina y evaluación del peligro sísmico. Actas XI Congreso Geológico Chileno 1:443–446

Paskoff R (1967) Antecedentes generales sobre la evolución del litoral del Norte de Chile durante el Plioceno y el Cuaternario. Boletín Sociedad De Geógrafos Chile 3:3–4

Paskoff R (1970) Le Chili Semiaride: Recherches, Géomorphologiques. Ed. Biscaye Fréres, Bordeaux, 420 pp

Pineda G, Calderón M (2008) Geología del área Monte Patria-El Maqui, región de Coquimbo, Escala 1:100 000. Carta Geológica de Chile, Serie Geología Básica, n.116, SERNAGEOMIN: 44 h. Santiago

Potter NJ (1972) Ice-cored rock glacier, Galena creek Northern Absaroka mountains, Wyoming. Geol Soc Am Bull 83(10):3025–3028

Potter N Jr, Steig EJ, Clark DH, Speece MA, Clark GT, Updike AB (1998) Galena Creek rock glacier revisited—New observations on an old controversy. Geogr Ann Ser B 80(3–4):251–265

Riquelme R, Rojas C, Aguilar G, Flores P (2011) Late Pleistocene—Early Holocene paraglacial and fluvial sediment history in the Turbio valley, semiarid Chilean Andes. Quat Res 75(1):166–175

Rivano S, Sepúlveda P (1991) Hoja Illapel, Región de Coquimbo. Carta Geológica de Chile, Sernageomin, Santiago, Chile

Rodríguez MP, Aguilar G, Urresty C, Charrier R (2015) Neogene landscape evolution in the Andes of north-central Chile between 28 and 32° S: interplay between tectonic and erosional processes. Geol Soc London Spec Publ 399(1):419–446

Santibañez F (1997) Tendencias seculares de la precipitación en Chile. In: Soto G, Ulloa F (eds) Diagnóstico Climático de la Desertificación en Chile. CONAF, vol 31, La Serena, Chile, pp 24–31

Soto MV, Ferrando F, Viera R (2002) Características geomorfológicas de un sistema de glaciares rocosos y de su cuenca de sustentación en Chile semiárido. Investigaciones Geográficas, Chile 36:1–16

Trombotto D, Wainstein P, Arenson LU (2014) Guía Terminológica de la Geocriología Sudamericana. Vazquez Mazzini Editores, Buenos Aires

Vuille M, Milana JP (2007) High-latitude forcing of regional aridification along the subtropical west coast of South America. Geophys Res Lett 34:L23703

Wahrhaftig C, Cox A (1959) Rock glacier in the Alaska range. Geol Soc Am Bull 70(4):383–436

Wayne WJ (1981) Ice segregation as an origin for lenses of non-glacial ice in "ice-cemented" rock glaciers. J Glaciol 27:506–510

Whalley WB (1974) Rock glaciers and their formation as part of a glacier debris-transport system. Geographic Papers, No. 27, University Reading, England

Whalley WB, Azizi F (1994) Models of flow of rock glaciers: analysis, critique and a possible test. Permafrost Periglac Process 5:37–51

Yáñez GA, Ranero CR, Huene R, Díaz J (2001) Magnetic anomaly interpretation across the southern central Andes (32–34 S): the role of the Juan Fernández Ridge in the late Tertiary evolution of the margin. J Geophys Res: Solid Earth 106(B4):6325–6345

Zech R, Kull C, Veit H (2006) Late Quaternary glacial history in the Encierro Valley, Northern Chile (29°S), deduced from 10Be surface exposure dating. Palaeogeogr Palaeoclimatol Palaeoecol 234(2–4):277–286

Zech R, Kull R, Kubik PW, Veit H (2007) Exposure dating of Late Glacial and pre-LGM moraines in the Cordon de Doña Rosa, Northern/Central Chile (31°S). Climate past Discuss 3:1–14

Zech R, May J-H, Kull C, Ilgner J, Kubik PW, Veit H (2008) Timing of the late Quaternary glaciations in the Andes from 15 to 40°S. J Quat Sci 23(6–7):635–647

Chapter 4
Timing and Mechanisms of Fluvial Responses to Rapid Land Use Changes in the Surrounding Highlands of the Pantanal Wetland (Brazil): The Case of Taquari River

Fabiano do Nascimento Pupim, Stefania Cristino de Oliveira, Beatriz Silva Fernandes, Ana Luiza Faria de Souza, Renan Cassimiro Brito, and Eder Renato Merino

Abstract Since the 1970s, the intensification of anthropogenic activities in the surrounding plateau has caused changes in surface processes that directly influence the fluvial dynamics of the Pantanal. As a consequence of these changes, the intensification of avulsion processes and increases in the magnitude of floods in the lowlands is observed, causing severe environmental and socioeconomic damage. Therefore, considering the ecological importance of the Pantanal and the changes that the basin has been suffering, there is a need to understand past and current conditions on fluvial behavior and how anthropogenic impacts on the plateaus have affected the lowlands. Here we propose to evaluate the response of the Taquari River to the human effects

F. N. Pupim (✉) · B. S. Fernandes · A. L. F. de Souza · R. C. Brito
Universidade Federal de São Paulo (Unifesp), Departamento de Ciências Ambientais, Rua São Nicolau 210-5 Andar, 09913-030 Diadema, São Paulo, Brazil
e-mail: f.pupim@unifesp.br

B. S. Fernandes
e-mail: beatriz.fernandes28@unifesp.br

A. L. F. de Souza
e-mail: ana.faria@unifesp.br

R. C. Brito
e-mail: renancb@usp.br

F. N. Pupim · S. C. de Oliveira · R. C. Brito
Universidade de São Paulo (USP), Instituto de Geociências, São Paulo, Brazil
e-mail: steoliveira@usp.br

E. R. Merino
Universidade de Brasília (UnB), Departamento de Geografia, Campus Universitário Darcy Ribeiro, GEA ICC Norte 900, Brasília, Distrito Federal, Brazil
e-mail: eder.merino@unb.br

A. Coronato and G. B. Alves (eds.), *Latin American Geomorphology*, The Latin American Studies Book Series, https://doi.org/10.1007/978-3-031-55178-9_4

on its upper course in the last 50 years. Our approach uses satellite images to estimate suspended sediment concentration and quantify morphological changes in the channel. The results obtained demonstrate that the impacts resulting from deforestation in the plateau interfere with the water discharge and sediment load along the river course. With the increase in cultivation and pasture and suppression of natural vegetation, there was an increase in production and sediment transport (soil erosion) in the basin, altering the river's dynamic, structure, and functioning. Thus, the response of water discharge and suspended sediments was rapid and synchronous with land use changes. However, the river bar response time was slower, with the most significant increase in bars occurring around 10–20 years after the peak of deforestation rates. Changes in channel width and migration rates were insignificant, suggesting that these parameters have greater resilience or longer response time. The information obtained here provides new data to predict anthropogenic effects on the river system and support plans for conservation.

Keywords River channel change · Sediment load · Deforestation · Anthropocene · Landsat time series

4.1 Introduction

Wetlands are environments periodically or continuously inundated by shallow water or soaked soils. They currently cover around 7% of the global land surface (Lehner and Döll 2004), with a variety of coastal and inland areas that are transitional between fully terrestrial and fully aquatic ecosystems (RAMSAR 2018). Wetlands have unique and complex geomorphological, hydrological, and biogeochemical behavior, which provide a wide range of economic and ecological services, climate regulation and maintain high biodiversity (Keddy et al. 2009; Hu et al. 2017; Finlayson et al. 2018). Yet, despite their importance, many wetlands have been threatened due to land degradation and overexploitation inside and on the highlands watersheds that drain into these areas (Davidson 2014; Hu et al. 2017).

Located in the center of South America, the Pantanal is one of the largest (~150.000 km^2) and most important freshwater wetlands on Earth (Alho 2005; Fig. 4.1). The anatomy and functioning of this wetland are strongly related to the water discharge regime and sediment load of rivers that comprise the Upper Paraguay River Basin (UPRB), which erode and transport of solutes (i.e., sediments and nutrients) from the uplands to the lowlands (Assine et al. 2015; Bergier and Assine 2022). In a millennial timescale, variations in flood frequency and magnitude, sediment delivery, and river morphology appear to be driven primarily by precipitation changes but influenced by land cover in the plateaus surrounding the Pantanal (McGlue et al. 2012; Pupim et al. 2017). However, inter-annual shifts in fluvial parameters are frequently related to changes in land use, dam operation, mining, and other human activities that cause variations in water and sediment supply along the catchment with significant effects in the lowlands (Gregory 2006; Wohl 2015; Best 2019).

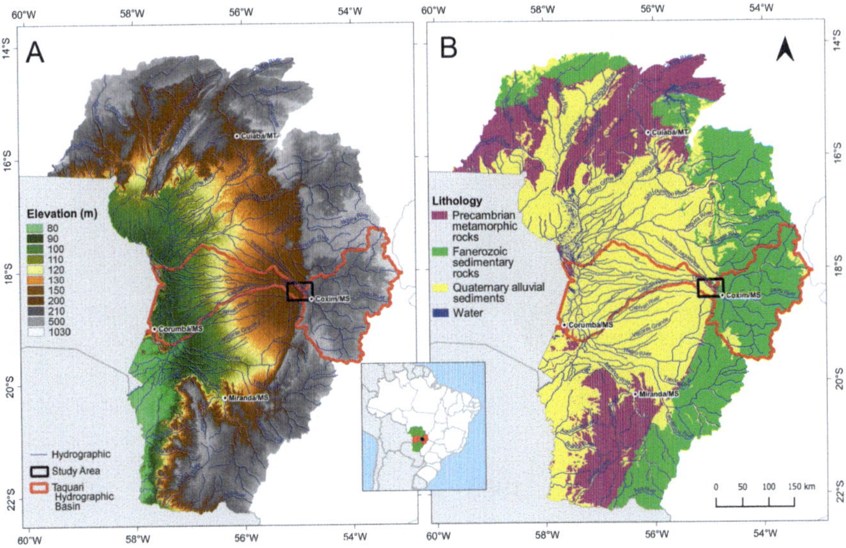

Fig. 4.1 Location of the upper Paraguay River Basin (UPRB) highlighted in red for the upper Taquari Basin (UTB). **a** Digital elevation model (data from SRTM, 90 m) showing the plateaus and lowlands (Pantanal). **b** Main geological units of the UPRB. Data from IBGE (2019)

In the UPRB, the most notable anthropogenic impact on the water and sediment regime is due to the deforestation of 61% of the native vegetation in the uplands to convert into pastures and croplands since the 1970s (Silva et al. 2005; Mapbiomas 2021), which had significant effects on the hydrological, sedimentological, biogeochemical, and ecological aspects of the lowlands (Bergier 2013; Roque et al. 2016). Land use changes are particularly evident in the Upper Taquari Basin (UTB), one of the most important rivers that feed the Pantanal wetland, where the increase of deforested areas was from 6 to 65% from 1976 to 2020 (Silva et al. 2005; Mapbiomas 2021; Fig. 4.2). Many studies suggest that the main consequences of this continuous land use change without soil conservation are the increase in runoff and soil erosion on the plateaus and, consequently, increasing the sedimentation in the middle and lower Taquari channel and its floodplains (Galdino et al. 2005; Makaske et al. 2012). However, such discussions are hindered by sparse and short-time monitoring programs on sediment load and ecological indicators from the riverine system. Remote sensing techniques using the long-term Lansat image collection can fulfill this knowledge gap since it has been successfully used to estimate long-term changes in biophysical parameters from rivers (e.g., Latrubesse et al. 2009; Montanher et al. 2014; Dethier et al. 2020).

The main aim of this study is to investigate the responses of the Taquari River due to environmental changes during the last 50 years. Here, we assume dams and mining are not significant factors in a watershed spatial scale because they are restricted to a few small tributaries of the UTB. Therefore, we hypothesized that inter-annual changes in hydrological, sedimentological, and morphological parameters of the middle and

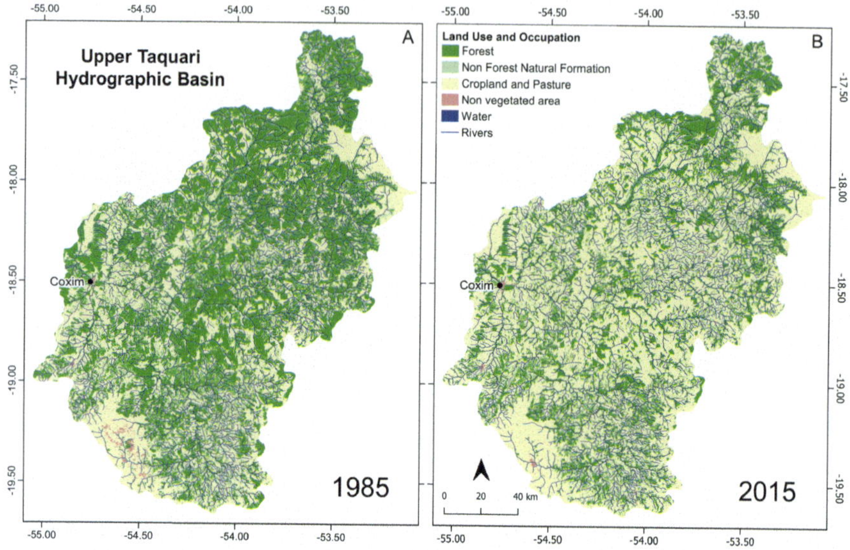

Fig. 4.2 Land use at upper Taquari Basin for **a** 1985 and **b** 2015 year (data from MapBiomas 2021). The reduction of forested areas (in green) and the increase of cropland and pasture areas (yellow) are apparent

lower Taquari River are related to changes in precipitation patterns or land use in the plateau terrains. We used images from the long Landsat temporal time series to quantify the changes in fluvial sand bars, channel width, lateral migration rates, and changes in the suspended sediment concentration. The data presented here allow for an estimation of the time and magnitude of the fluvial responses to changes in land use, providing new information on the connection between the UTB plateaus and the Pantanal wetlands. This new understanding can inform managers and decision-makers to promote an economic agenda based on sustainable development concepts that allow conserving the rivers and wetlands of the Pantanal.

4.1.1 Physical Settings

The Upper Paraguay River Basin (UPRB) is a vast physiographic mosaic located at the center of the South American continent, where the Pantanal is the most known landscape. The total area of the UPRB is about 600.000 km^2 within the Mato Grosso and Mato Grosso do Sul states, central-west Brazil, and reaching territories of Bolivia and Paraguay (ANA 2004; Fig. 4.1). The geomorphological and geological setting of the catchment is composed of plateaus that consist of calcareous formations in the north (Serra das Araras) and in the south (Serra da Bodoquena), basalts of the Serra Geral Formation toward the south-east, sandstone formations in the eastern border, and crystalline rocks composing the basement (Assine et al. 2015). At the center of the UPRB, the Pantanal is an active sedimentary basin continually filled by Cenozoic sediment transported by the Paraguay trunk river and its right margin main tributaries. This system drains the water and part of the sediment load received from the surrounding catchment areas of the UPRB toward the Pantanal Basin, creating a setting of megafans and interfan systems in which the trunk plain is the regional base level that controls the flood dynamics of one of the world's largest wetlands (Assine et al. 2015).

Among the largest rivers of the UPRB, the Taquari River is one of the most important tributaries. The river is approximately 800 km long, with headwaters in the highlands between the Saudade and Maracaju plateaus in Mato Grosso do Sul. The high erosion rates in the UTB catchment area are responsible for 72% of the total amount of suspended sediment delivered to the Pantanal lowlands (Risso et al. 1997). This system is responsible for the origin and modeling of the Taquari megafan, a geomorphological unit with a circular shape and 50.000 km^2 of area (Assine 2005). The megafan is radially divided into two regions by the Taquari Rivers that present E–W flow direction. The northern zone is locally known as Paiaguás, while the southern border is called Nhecolândia, one of the most iconic sub-regions of the Pantanal. In this unique lacustrine system, more than 10.000 saline-alkaline and freshwater lakes and ponds coexist (Zani et al. 2012). In addition, both zones present numerous preserved paleochannels that spread from the entrance of the Taquari River in the Pantanal Basin towards all directions of the megafan (Zani et al. 2012). This intense hydro-sedimentological dynamic is controlled by the natural characteristics of the UTB, including its climate, geological, and physiographic settings.

According to Köppen's classification, the climate of UTB is tropical (Aw) (Alvares et al. 2013), with well-defined dry (June–August) and wet (December–March) seasons. The mean annual temperatures range from 21–30 °C in the catchment to 21–31 °C in the Pantanal Basin. The South American Summer Monsoon directly influences the region by causing heavy rainfall events during the austral summer (Zhou and Lau 1998). While the maximum rainfall is nearly 1800 mm in the catchment, within the Pantanal is ~1500 mm (Funk et al. 2015), the last one responsible for the negative water balance, with evapotranspiration rates higher than the rainfall (McGlue et al. 2011). The monsoon precipitation provides a significant water supply resulting in a high amount of discharge in the rivers, directly affecting the annual flood pulse and sediment connectivity of the Pantanal wetland.

The UTB terrains consist of sedimentary and igneous rocks from the Paraná Sedimentary Basin, mainly sandstones from the Pirambóia and Botucatu formations, and underneath Precambrian metamorphic and granitoid rocks in the basement (IBGE 2019; Fig. 4.1). The Taquari megafan is primarily composed of sands with Pleistocene and Holocene age (Assine 2005). Steppe slopes in the UTB are represented by Leptsols and Arenosols, which are very susceptible to erosion, especially the development of gullies (Galdino et al. 2005). Farralsols and Lixisols are associated with flat and undulating terrains, and the hydromorphic soils are constrained within the fluvial plains and entrenched valleys (Galdino et al. 2005).

4.2 Data and Methods

4.2.1 Data Collection and Processing

The central database used in this study is satellite images from the sensors Thematic Mapper (TM), Enhanced Thematic Mapper (ETM), and Operational Land Imager (OLI) of the Landsat Program. These sensors capture images in six spectral bands from visible to short-wave infrared, with a spatial resolution of 30 m. We processed and acquired the images using the Google Earth Engine platform (2019) to map morphological parameters of the river channel and estimate SSC between 1985 and 2021. In addition, we used the DSGTools supporting features on Quantum GIS http://plugins.qgis.org/plugins/DsgTools) to recover morphological parameters based on topographic maps from 1970.

Rainfall and hydrological data (water elevation, flow, and transported suspended sediment load) were collected by the Brazilian National Water Agency website (ANA—www.ana.gov.br/sar/national-water-agency). From the hydro-sedimentological data, it was possible to analyze and compare changes in fluvial dynamics over time and their relationships with anthropic impacts developed in the UTB. It provided information to estimate the concentration of suspended sediments. Additional rainfall data was acquired from the Climate Hazards Group InfraRed Precipitation with Station data (CHIRPS, Funk et al. 2015).

Data on land use changes in 1976 and 1984 were obtained from publications by Embrapa—Empresa Brasileira de Pesquisa Agropecuária (SILVA et al. 2005), which presents data on changes in land use based on the interpretation of aerial photographs for the period between the 1970 and 2000. The quantification of land use change between 1995 and 2020 using data from the MapBiomas Project (mapbiomas.org), which uses cloud processing and automated classifiers developed and operated using the Google Earth Engine platform. The organization of these data occurred through the cut of shapefiles with annual periodicity referring only to the area of the Taquari river basin upstream of Coxim city. The Forest and Natural Forest Formation category was grouped in a category called Natural; the other categories were grouped in the Anthropic Use category. The reduction in the Forest categories was interpreted as an increase in deforestation.

4.2.2 Estimating the Suspended Sediment Concentration

The suspended sediment concentration (SSC) is an essential component in the quality of river waters that influences the physical, chemical, biological, and geomorphological processes of the channel (Schumm 2007). SSC monitoring is a fundamental tool in identifying areas with unbalanced erosion processes in hydrographic basins, planning and operating reservoirs or hydroelectric plants, and recovering degraded aquatic environments (Markert et al. 2018). However, the sampling and spatial frequency are generally low due to high costs, technical restrictions, and collections in places of difficult access. In this context, considering the lack of data in the time series of SSC data in the UTB rivers (e.g., ANA), the estimation through in situ point averages and historical series of orbital data available for free is a valuable tool for the knowledge of the fluvial dynamics and environmental monitoring.

The method for estimating the SSC from remote sensors is based on the premise that the spectral reflectance varies as a function of the SSC. Hence, as the SSC in water increases, there is an increase in electromagnetic radiation (EMR) backscatter and, consequently, in the reflectance of the water body (Montanher et al. 2014). Therefore, regression analysis is usually applied to express the relationship between reflected ERM and SSC, in which the dependent variable (SSC) can be estimated as a function of one or more independent variables (Dethier et al. 2020).

The chosen gauge station was Coxim, located near the boundary between the plateau and the lowlands. After obtaining and selecting the data, we found that the in situ samples presented a temporal coverage from 2005 to 2019, totaling 33 samples, showing an SSC variation from 0 to 1254 mg L^{-1}.

The choice of the Landsat System to collect the orbital data was due to its wide temporal range. Images were used from 1984 with the launch of Landsat 5 TM (1984–2012) to the present day with Landsat 7 ETM+ (1999–2020). In addition, it was decided on the USGS Reflectance Surface product that has the scattering and radiation absorption effects in the atmosphere minimized through a radiative transfer model with the LEDAPS algorithm (USGS 2020). In total, 499 images were collected, and then a stack of all images in the Landsat 5 and 7 SR record.

Images were filtered, and criteria were used to eliminate waste that could contaminate the selection of water, according to Dethier et al. (2020). The Landsat Reflectance Surface product includes land cover classification and quality assurance codes for each pixel in the "pixel_qa" band. USGS classifications identify clouds and cloud shadows, so we use these classifications to remove pixels with these disturbances (USGS 2020). In addition, complementary reflectance thresholds were added for a more secure approach to classification for cloud removal, cloud edges, haze, and water and land pixel distinction.

For each image with the respective field collection, a sampling area was defined within a radius of 30 m close to the gauge station of Coxim. From this sample area, the selected pixels were averaged. In this methodological step, sampling in places with shallow water with bottom effect and spectral mixing with banks and sandbanks was avoided.

After collecting the in situ and orbital data, data calibration was performed in RStudio, which uses a programming language to develop graphs and statistical calculations (RStudio Team 2021). The generation of models requires that spectral reflectance measurements of water be used almost simultaneously with SSC measurements at ANA gauge stations. It was assumed that a few days of variation in the field collection date with imaging would not drastically alter the natural variation in the sediment load within the water column (Montanher et al. 2014). In the present study, 12 images were selected within ±3 lagged days of in situ collections of SSC.

The construction of the model was based on a simplified approach using multiple linear regression with continuous independent variables being the ratio between spectral bands and categorical independent variables, the sensors of the Landsat 5 TM and 7 ETM+ satellites. The LOOCV (leave-one-out cross-validation) cross-validation was performed for validation purposes. Given the saturation effect of the spectral response of the optically active components in the water column as their concentration increases and a nonlinear correction is obtained, the data underwent a natural logarithmic transformation (e.g., Cremon et al. 2019).

The choice of the most appropriate band rate for the model was based on the compilation of band rates and indices used in other works that showed promising results (Montanher et al. 2014; Cremon et al. 2019). According to the coefficient of determination (R^2) and the square root of the mean error (RMSE), the band ratio $B4 + B3/B2 + B1$ was adopted to generate the models for estimating suspended sediments. Finally, the model was adjusted by a categorical predictor variable whose value refers to the coefficient of the Landsat sensor of the images, resulting in Eq. 4.1. The performance of the model in terms of R^2 and RMSE determination was respectively 0.58 and 0.29.

$$\mathrm{SSC} = -0.77 + 1.30 \times ((B4 + B3)/(B2 + B1)) + 0.15(\mathrm{Landsat\,7}) \tag{4.1}$$

4.2.3 Detection of Morphological Changes

Landsat images from 1985, 1995, 2005 (TM), and 2015 (OLI) were used to map and quantify changes in fluvial bars, channel width, and migration rates. In addition, data from topographic maps were used to map river morphology from 1970. We selected images from dry periods (June–August) with low river water levels to minimize the influence of seasonal variations in water level in estimating the morphological parameters, especially the area of fluvial bars and channel width. The images were processed using a false-color composition with bands 5(R), 4(G), and 3(B) to the years 1985–2005 (TM sensor), bands 6(R), 5(G), and 4(B) to 2015 (OLI sensor). The identification and mapping of the channel margins and fluvial bars were performed by visual interpretation of each scene. The mapping area was from Coxim city to 100 km downstream in the Taquari River (Fig. 4.3a). The analysis was conducted in two scales, one that takes the entire area and another that takes segments of 10 km along the river. This division aims to evaluate changes in the detailed scale and identify possible local controls that affect channel dynamics.

We identified fluvial bars as depositional landforms that emerge into the channel (Fig. 4.3). Five indicators were used: the number of river bars; the area occupied by river bars; the size of bars (small, medium, and large); the type of bars according to their position in the channel (lateral and mid-channel bars) and classify them into vegetated and non-vegetated bars (Latrubesse et al. 2009). The classes of bar size were defined based on the average and standard deviation of all mapped bars, which

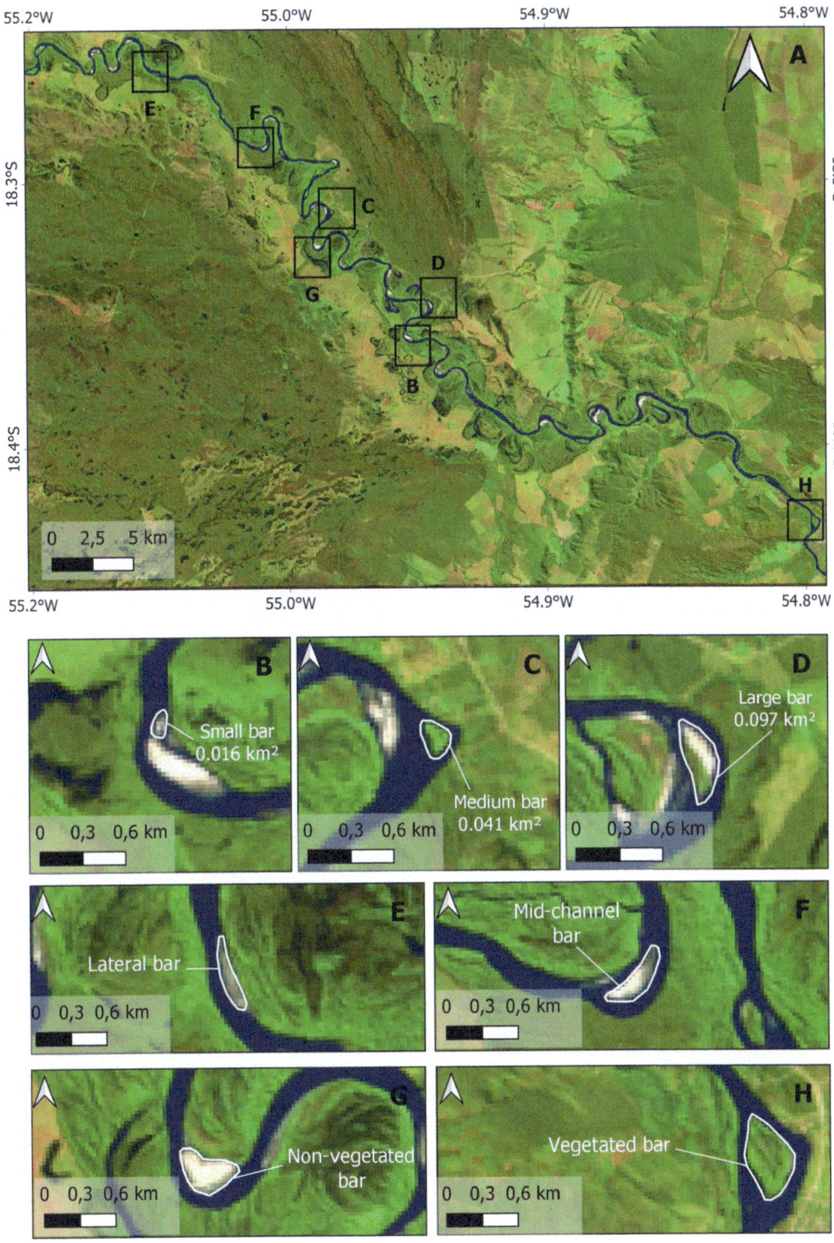

Fig. 4.3 The studied channel reach (**a**) and types of fluvial bars mapped from Landsat images in the middle Taquari River (**b**–**h**). The fluvial bars were classified according to their size (small, medium, and large), their position in the channel (lateral and mid-channel bars), and vegetated and non-vegetated bars. See localization of A in Fig. 4.1

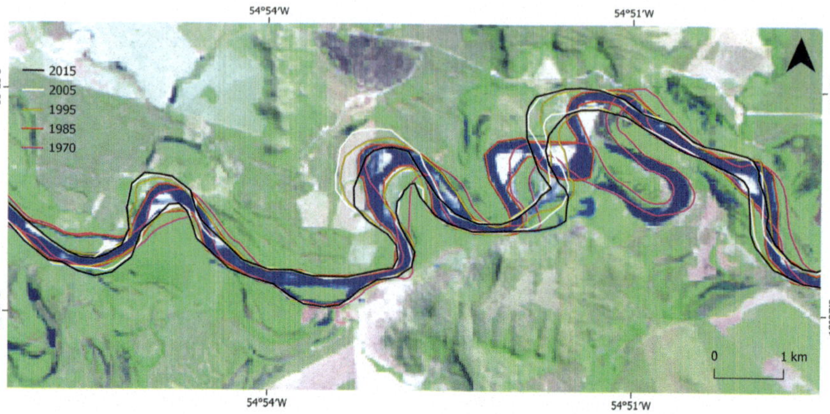

Fig. 4.4 Defining riverbank to estimate changes in channel width and migration rates from 1970 to 2015

were the small bars from 0.001 to 0.02 km^2, medium bars from 0.021 to 0.05 km^2, and large bars from 0.051 to 0.78 km^2.

The estimation of the channel width and migration rates starts with the visual mapping of both river margins from 1970 to 2015 (Fig. 4.4). The average channel width from each reach of 10 km was measured using the area of the channel divided by the length of the reach (i.e., 10 km).

For channel migration rates, based on both margins, a channel centerline was generated automatically for the tool Collapse Dual Lines to Centerlines in the ArcGIS 10.6.1 software. Next, the centerline of each year was superposed to the following year's line; in other words, the 1970 line was compared to the 1985 line, the 1985 line was compared to 1995, and so until 2015. Finally, a polygon was generated from the two central lines to quantify the area of the river displacement in all sections, and Eq. 4.2 was applied to calculate the mean channel migration rates.

$$\underline{C} = C_1 + C_2 \tag{4.2}$$

Lateral migration rates = Polygon area of $\underline{C}$

C_1 = channel length in the previous decade; C_2 = channel length in the following decade.

4.3 Temporal Pattern of Precipitation, Land Use, and Fluvial Data

4.3.1 Regional Precipitation and Land Use Changes

To investigate the possible drivers of fluvial changes in UTB, we create a time series for two predictor variables: a decadal land use change series from 1976 to 2020 (Fig. 4.5), a monthly average rainfall series from 1966 to 2021 (Fig. 4.6a).

The land use change data show that 66% of native vegetation (Cerrado) in the plateau of Taquari river has been converted into anthropic areas until 2020 (Fig. 4.5). Deforestation was extremely fast from 1976 to 1994 when 56% of the natural vegetation was suppressed. From 1995 to 2020, deforestation rates decreased, increasing by 10% of new anthropic areas in the UTB. This land use evolution occurred due to intensive deforestation and conversion of natural vegetation to pasture for beef production (~80%) and croplands for soy, corn, and cotton (~19%) (Mapbiomas 2021). Urbanized areas, silviculture, and other uses are less than 1% of the total anthropic areas.

Monthly rainfall showed high intra-annual variability, with wet summer (Nov–Jan) and dry winter (Jun–Ago). However, the long-term time series showed no consistent increased or decreased trends, which means that the amount of rainfall over the UTB has not significantly changed over the past 50 years (Fig. 4.6a). This observation agrees with Bergier et al. (2018). They analyzed a long-time series from 1926 to 2016 and suggested that the decadal rainfall trend was not statistically significant for all seasons (autumn, winter, spring, and summer) in the Pantanal. However, they also showed a decrease in the mean number of rainy days but an increase in the amount of rain per day, which means that rainfall becomes concentrated in time, i.e., an increase in extreme rainfall events (Bergier et al. 2018).

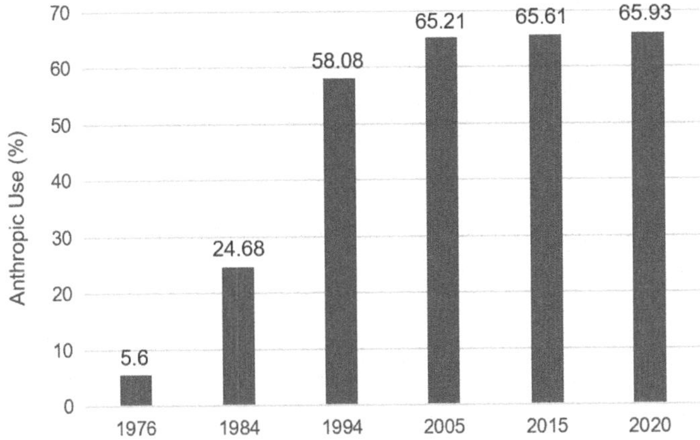

Fig. 4.5 Land use changes in the upper Taquari River. From 1976 to 2020, 66% of native vegetation was converted into anthropogenic areas. Data from Silva et al. (2005), MapBiomas (2021)

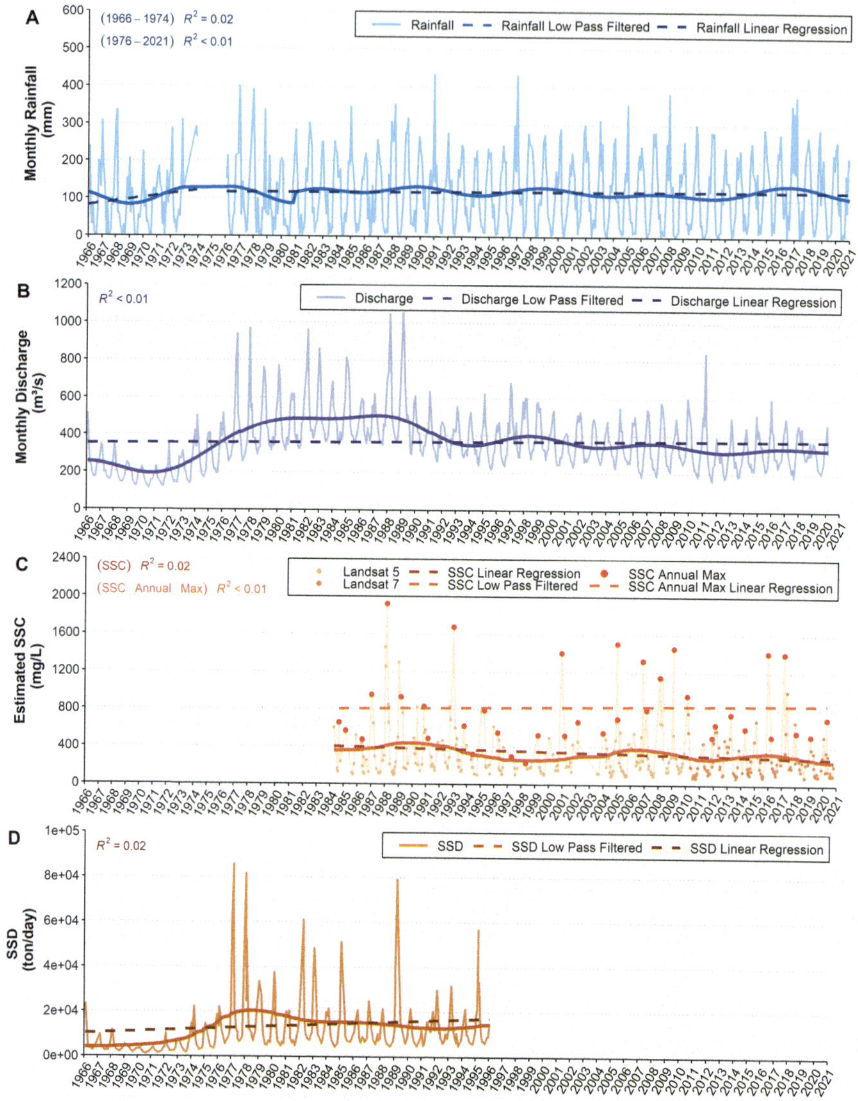

Fig. 4.6 Time series from 1966 to 2021 showing rainfall variability, water discharge, and suspended sediment concentration in the Taquari River. **a** Monthly average rainfall series (data from ANA—www.ana.gov.br/sar/national-water-agency and from CHIRPS—Funk et al. 2015). **b** Monthly average water discharge series (data from ANA). **c** SSC series estimated from Landsat images (this study). Data of the modeled total suspended sediment discharge were provided by Barbedo (2003)

4.3.2 Water Discharge and Suspended Sediments Load

We generated a time series dataset of the mean annual water discharge and estimated SSC of the UTB upstream entrance into the Pantanal wetland, using gauge station data (Fig. 4.6b) and a multiple regression model from Landsat images (Fig. 4.6c), respectively. Discharge and SSC exhibit strong inter-annual variation, and their evolution and trends appear to be correlated from 1966 to 2021. Although we could not estimate SSC before 1985, we used previous data of suspended sediment discharge modeled by Barbedo (2003) for the period between 1966 and 1995 (Fig. 4.6d).

The water discharge series did not show a systematic trend from 1966 to 2021 (Fig. 4.6b). In contrast, we observed two sub periods along the series. First, the discharge increases (~150%) systematically from the end of 1971–1980, followed by a slight increase until 1988. From 1989 to 1992, an abrupt change was observed, and the discharge decreased by around 20%. Second, a small but clear downward trend is observed in water discharge from 1993 to 2021. The intra-annual variability of mean monthly discharge indicates a one-month delay regarding rainfall, with the highest discharge occurring from Dec. to Feb. and the lowest discharge from Jul. to Sep.

The composed suspended sediment time series, using previous data (1966–1995) and our new estimated SSC data (1985–2021), suggest a good correlation with water discharge, with two sub periods along the series from 1966 to 2021 (Fig. 4.6c, d). Modeled suspended sediment discharge shows a rapid increase from 1971 to 1978, then relative stability until 1995. The SSC data shows a slight linear decrease trend of the mean values along the time series (1985–2021).

4.3.3 Fluvial Bars and Channel Morphology

The mapping and quantification of the morphological indicators of the middle Taquari River, a reach 100 km downstream of Coxim city (Fig. 4.3a), demonstrate that there was an apparent increase in the quantity and area occupied by fluvial bars and a tendency to slightly decrease in the channel width and migration rate over the time series from 1970 to 2015. Results are summarized in Fig. 4.7 and will be detailed described below.

The data of fluvial bars indicate that the most significant changes occurred after 1995, with an increase of almost 3.5 times the quantity and area occupied by fluvial bars along the middle Taquari River (Fig. 4.7a). Furthermore, spatiotemporal variations were observed in the accumulation of fluvial bars along river reaches. The distribution of bars was relatively homogeneous in the period between 1970 and 1995, changing to a heterogeneous distribution pattern after 2005, from when the fluvial bars area was about twice as large in the reach of 0–60 km than in the reach 60–100 km downstream from Coxim (Fig. 4.7e).

In addition to changes in the area and quantity of fluvial bars, the data demonstrate changes in the size and position within the channel and the relation between non-vegetated and vegetated bars. After 1995, the number of small bars increased

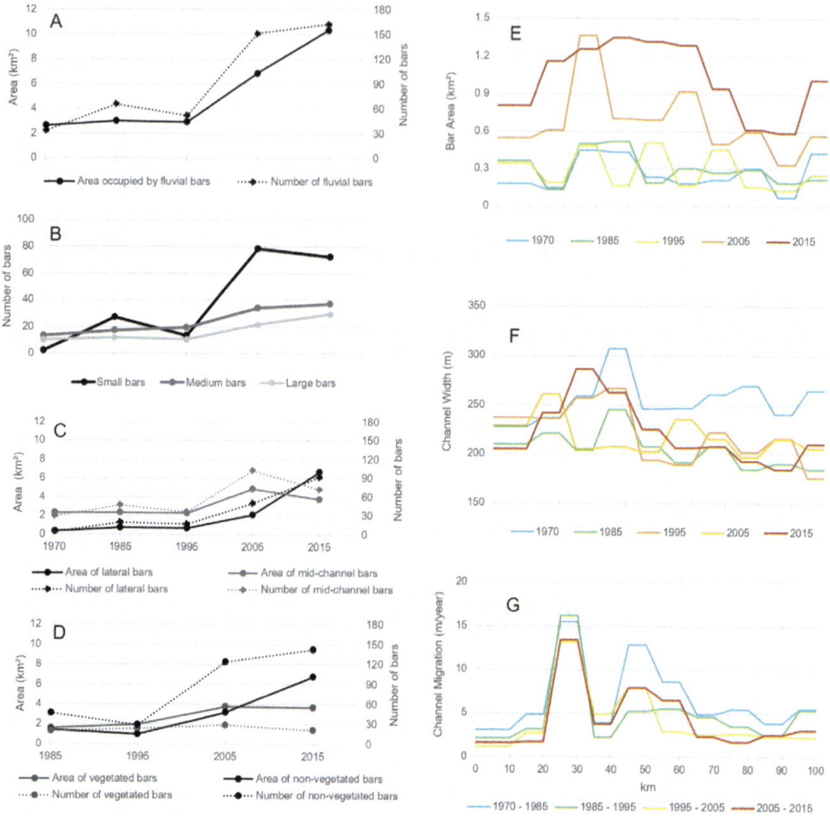

Fig. 4.7 Morphological changes in the Taquari River from 1970 to 2015. Fluvial bars were quantified by **a** area and the number of occurrences and classified as **b** size, **c** position in the channel, and **d** vegetated and non-vegetated. Longitudinal and temporal variations of the fluvial bar area (**e**), channel width (**f**), and channel migration rates **g** were also measured

about four times, while the medium and large bars increased about two and three times, respectively (Fig. 4.7b). The mid-channel and lateral bars had similar increases between 1995 and 2005, but with a change of trend between 2005 and 2015, when the lateral bars continued to increase in quantity and area, and the mid-channel bars suffered slight decreases in both parameters (Fig. 4.7c). Another quantified change was the significant increase in the number and area of non-vegetated bars (from 1.41 km^2 in 1985 to 6.71 km^2 in 2015) against apparent stability of vegetated bars for the period from 1985 to 2015 (Fig. 4.7d).

The channel width showed no-linear variations during the 50-year time series (Fig. 4.7f). A trend of channel narrowing downstream was recognized in all analyzed periods, with a prominent decrease after the 40–50 km reach, indicating water depletion from the channel to the floodplain. The maximum and minimum channel width averages were registered in the 1970 and 1985 time lapse, respectively, showing a channel narrowing of about 10%. From 1985 to 2015, the channel width greatly varies along the reach from 0 to 50 km downstream, but with a non-significant increase or decrease. In contrast, no significant width variability was observed in the reach from 50 to 100 km.

The channel migration rates slightly decreased in some lengths from 1970 to 1995 and show relative stability with low migration rates from 1995 to 2015 (Fig. 4.7g). From 1970 to 1985, the highest migration rates were observed from 30 to 40 km (15.5 m/y), with another peak in the length from 50 to 60 km (12.8 m/y). Likewise, maximum channel migration between the 1985 and 2015 periods was observed between 30 and 40 km, with rates from 16.2 to 13.4 m/y. Furthermore, the most common channel migration rate at all periods was from 2 to 7 m/y, and the minimum channel migration rate occurred in the most upstream reaches, ranging from 1.2 to 3 m/y.

4.4 Discussion

The analysis presented above demonstrates significant changes in water discharge, SSC, and channel morphology over the last 50 years in the Taquari River. However, channel width and migration rates do not show significant trends. In what follows, we discuss the results regarding driving forces, mechanisms, and response time for temporal variation of analyzed fluvial parameters.

By comparing our long-term fluvial dataset with precipitation and land use time series (predictor variables), we observe a clear intra-annual response of water discharge, SSC, and fluvial bars to land use change. The intensive conversion of natural vegetation to pastures and croplands in the uplands during the 1970s and 1980s is reflected rapidly in the substantial increase of water discharge and suspended sediment load to the lowlands (Figs. 4.5 and 4.6). Although, from 1990 to 2021, changes in land use were slow and gradual, reflecting in stabilization to a slight decrease in water discharge and SSC. Despite the lack of a significant long-term link between changes in the fluvial data with precipitation series, the magnitude of discharge and SSC appear to have been influenced by the decadal oscillation of precipitation. This influence was more prominent during the 1970s, with a positive correlation between precipitation and discharge increase. In the last two decades, when we can observe a coupled oscillation between those variables (Fig. 4.6). Those finds agree with local and global empirical studies. Recently, satellite image-based research highlighted that intensive land use change in the global south has increased erosion, with river-suspended sediment concentration on average 41 $\pm$ 7% greater than in the 1980s (Dethier et al. 2022). In the Taquari River and other rivers from UPRB, data from floodplain sediments and shorter timescale river monitoring suggest that changes in hydrology and sediment flux in the lowlands are primarily driven by changes in land use in the uplands, with the most critical period of changes occurring during the 1970–1980s (Godoy et al. 2002; Galdino et al. 2005; Bergier 2013).

We suggest a mechanism that, without vegetation cover and soil conservation practices, the first 20 years of land use changes would impact soil characteristics and surface hydrology, leading to a significant increase in water runoff (decrease infiltration), soil erosion, and sediment delivery from the slope to the channel. In contrast, the reduction of deforestation, advances in pasture management, and soil conservation practices would lead to a control of water runoff (increase infiltration) and soil erosion, resulting in the observed gradual decrease of river discharge and SSC. Similar responses and mechanisms are also reported worldwide in different spatial and temporal scales (e.g., Gregory 2006; Wohl 2015; Dethier et al. 2022).

The increase of fluvial bars area by 350% between 1995 and 2015 indicates an increase in significant bedload sediment, which appears to be linked with upland land use change that led to headwater erosion and downstream aggradation of the valley bottom. Our data also demonstrate a spatial heterogeneity of bars distribution, in which sediment accumulation (bar formation) occurs from upstream to downstream (Fig. 4.7). The mechanism that led to the increase of fluvial bars is the same described as changes in water discharge and SSC. However, the response time of fluvial bars seems slower than hydrologic variables, with a delay of 10–20 years concerning the most relevant period of land use change. The increase of fluvial bars between 1995 and 2015 reflects land use changes during the 1970s–1980s (Fig. 4.7). Latrubesse et al. (2009) and Suizu et al. (2022) also suggest that land use changes are the main driver of the fluvial bars increase in the Araguaia River in the last 50 years, with a similar delay of 10–20 years in the response time. This delayed response time could be due to sediment storage in the floodplains along the system since the detaching and transport of coarse grains (sand) are limited to flashier discharges. This sediment storage hypothesis is supported by data that indicate an increase of 500% in siltation on oxbow lakes in the Taquari River floodplain from 1950 to 1990 (Godoy et al. 2002). The similar response of the Araguaia and Taquari basins is of great interest to regional management since they are in a similar climate and land use setting, suggesting a temporal and spatial pattern for the fluvial responses to land use changes in rivers that drain the Brazil Central.

In addition to changes in the area and quantity of fluvial bars, we also observe a relative increase in bars characterized as small (area), lateral (position in the channel), and non-vegetated (Fig. 4.7). The predominance of non-vegetation bars suggests that sediment input is relatively recent and continuous, or the sediment residence time in the channel is short, limiting the vegetation colonization and long-term stabilization of the bars (Corenblit et al. 2020). This high and continuous sediment input reduces channel depth and sediment transport capacity; consequently, sediments are temporarily stabilized and emerge as non-vegetated fluvial bars, initially smaller in area. The formation of larger bars occurs due to the amalgamation of small bars, suggesting a continuous and high sediment input to the Taquari River. The increase of lateral bars about mid-channel bars appears to respond to high channel aggradation and annexation of large bars to the riverbank.

It is well documented that channel migration rates positively correlate with sediment loads (Constantine et al. 2014). The channel width is primarily driven by riverbank cohesion but is also influenced by water discharge and sediment load (Dunne and Jerolmack 2020). Despite the increase in water discharge and sediment loads (suspension and bed load), channel width and migration rates of the Taquari River slightly decreased from the 1970s to the following decades. These decrease in width and migration rates would respond to the increase in water discharge that promotes river incision. However, channel width and migration rates have stayed roughly constant after 1985. Mason and Mohring (2019), using high-resolution LIDAR images in the Trinity River (USA), suggest that on a yearly scale, the river bend widens and narrows in distinct steps. Following these modifications, the opposing tendency balances them out, resulting in a statistically steady-state or "equilibrium" channel width around the decadal scale. Broadly, for the Taquari River, we interpret that channel width and migration rates were not sensitive to flow and sediment load changes over a multi-decadal time scale, indicating a longer response time.

The above analysis offers a quantitative approach to investigating and monitoring changes in river flow and sediment regimes using remote sensing techniques. We demonstrate that converting natural forests to agropastoral lands led to a critical increase in water discharge and sediment load from the UTB to the lowland areas in Pantanal. However, the implications of the land use change in the uplands around the Pantanal have effects far beyond the immediate changes in flow (i.e., flood pulse) and sediment regime (i.e., siltation and water turbidity). An exciting implication of the increase in river discharge and sediment load is the acceleration channel/floodplain siltation and the frequency of river avulsion phenomena in the lower Taquari River (Taquari megafan), which has been documented two times in the last 50 years (Assine 2005). These processes affect the spatial arrangement of inundated areas, vegetation (Assine 2005; Ivory et al. 2019), and ecological processes and services (Louzada et al. 2020, 2021). Moreover, changes in hydro-sedimentological processes cause damage to aquatic habitats (i.e., type and connectivity) and decrease the abundance and diversity of benthic invertebrates (Wantzen et al. 2005), change biogeochemical cycle (Hamilton et al. 1997), and impact herpetofauna breeding and fish migration routes, for example (Alho and Sabino 2012). Another cumulative impact related to the increase of runoff, soil erosion, and sediment load is the increase of contaminants bound in sediments (e.g., pesticides and mercury), which have potentially catastrophic implications for vegetation, animals, and human health (Souza et al. 2020).

It is worth debating future scenarios for UTB because they may give insights into the perception of future trends related to fluvial response to land degradation and inform conservation and development plans in the UPRB and Pantanal wetlands. Broadly, our finds in the UTB agree with numerical models, which predict an increase of 100% in sediment load in the lowlands due to a decrease of 50% in natural vegetation in the watershed (Coulthard and Van De Wiel 2017). Moreover, based on the referenced numerical simulations, the reforestation of critical terrains in UTB could lead to significant decreases in sediment load to the Pantanal wetland.

4.5 Conclusions

Based on remote sensing and hydrologic data, we have directly quantified hydrological and sedimentological changes over 50 years in the Taquari River, the most important sediment feeder of the Pantanal wetland. Our data reinforce that land use changes (i.e., deforestation) in the upland areas are the primary driver of hydrological, sedimentological, and morphological changes in the lowland sector of the river.

The exponential increase in the conversion of natural vegetation to croplands and pastures in the 1970s and 1980s lead to an increase in runoff and sediment production (soil erosion) on the slopes, consequently impacting river dynamics, structure, and functioning. The response of water discharge and suspended sediments was fast and synchronous with changes in land use. However, the response time of the fluvial bar was slow, and the highest fluvial bars increased around 10–20 years after the peak of deforestation rates. Changes in channel width and migration rates were not significant, suggesting that these parameters have higher resilience or longer response time. Finally, we highlight that, despite the relative stabilization of the hydrological and sedimentological characteristics of the Taquari River, they did not return to initial conditions as recorded in the 1970s.

It is clear that the land use policy has threatened the Pantanal wetland environmental balance because the last 50 years of deforestation increased the river discharge and sediment load from its surrounding uplands. In the future, however, the climate changes and projected dams can synergically promote new cumulative changes in the water-sediment cascade. Therefore, if applied on a detailed scale, we suggest that the approach presented here can help identify the disequilibrium in sediment supply and areas prone to erosion, then indicate candidate sites for reforestation projects.

Acknowledgements The authors thank Ivan Bergier (EMBRAPA) for the review and suggestions. This research was supported by several individual grants, such as FP (#302411/2018-6) is a Brazilian National Council for Scientific and Technological Development—CNPq/Brazil research fellow. SO is a graduate scholarship from CAPES. BF has an undergraduate scholarship from the Sao Paulo Research Foundation—FAPESP (grant # 2020/07432-7).

References

Alho CJR (2005) The Pantanal. In: Fraser LH, Keddy PA (eds) The world's largest Wetlands—Ecology and conservation. Cambridge University Press, New York, pp 203–271

Alho CJ, Sabino J (2012) Seasonal Pantanal flood pulse: implications for biodiversity. Oecol Aust 16(4):958–978

Alvares CA, Stape JL, Sentelhas PC, Gonçalves JDM, Sparovek G (2013) Köppen's climate classification map for Brazil. Meteorologische Zeitschrift 22(6):711–728

ANA-Agência Nacional de Águas (2004) Implementação de Práticas de Gerenciamento Integrado de Bacia Hidrográfica para o Pantanal e Bacia do Alto Paraguai ANA/GEF/PNUMA/OEA., Brasília

Assine ML (2005) River avulsions on the Taquari megafan, Pantanal wetland, Brazil. Geomorphology 70(3–4):357–371

Assine ML, Merino ER, Pupim FDN, Macedo HDA, Santos MGMD (2015) The quaternary alluvial systems tract of the Pantanal Basin, Brazil. Braz J Geol 45(3):475–489

Barbedo AGA (2003) Estudo hidrossedimentológico na Bacia do Alto Paraguai, Pantanal. https://repositorih.ufms.br/handle/123456789/1474

Bergier I (2013) Effects of highland land-use over lowlands of the Brazilian Pantanal. Sci Total Environ 463–464:1060–1066

Bergier I, Assine ML, McGlue MM, Alho CJR, Silva A, Guerreiro RL, Carvalho JC (2018) Amazon rainforest modulation of water security in the Pantanal wetland. Sci Total Environ 619–620:1116–1125
Bergier I, Assine ML (2022) Functional fluvial landforms of the Pantanal: hydrologic trends and responses to climate changes. J S Am Earth Sci 119:103977
Best J (2019) Anthropogenic stresses on the world's big rivers. Nat Geosci 12(1):7–21
Constantine JA, Dunne T, Ahmed J, Legleiter C, Lazarus ED (2014) Sediment supply as a driver of river meandering and floodplain evolution in the Amazon Basin. Nat Geosci 7(12):899–903
Corenblit D, Vautier F, González E, Steiger J (2020) Formation and dynamics of vegetated fluvial landforms follow the biogeomorphological succession model in a channelized river. Earth Surf Proc Land 45(9):2020–2035
Coulthard TJ, Van De Wiel MJ (2017) Modeling long term basin scale sediment connectivity, driven by spatial land use changes. Geomorphology 277:265–281
Cremon ÉH, da Silva AMS, Montanher OC (2019) Estimating the suspended sediment concentration from TM/Landsat-5 images for the Araguaia River—Brazil. Remote Sens Lett 11(1):47–56
Davidson NC (2014) How much wetland has the world lost? Long-term and recent trends in global wetland area. Mar Freshw Res 65(10):934–941
Dethier EN, Renshaw CE, Magilligan FJ (2020) Toward improved accuracy of remote sensing approaches for quantifying suspended sediment: implications for suspended-sediment monitoring. J Geophys Res: Earth Surf 125(7):e2019JF005033
Dethier EN, Renshaw CE, Magilligan FJ (2022) Rapid changes to global river suspended sediment flux by humans. Science 376(6600):1447–1452
Dunne KBJ, Jerolmack DJ (2020) What sets river width? Sci Adv 6(41):eabc1505
Finlayson CM, Milton GR, Prentice RC (2018) Wetland types and distribution. In: Finlayson CM, Milton GR, Prentice RC, Davidson NC (eds) The Wetland book: II: distribution, description, and conservation. Springer, Netherlands, Dordrecht, pp 19–35
Funk C, Peterson P, Landsfeld M, Pedreros D, Verdin J, Shukla S, Husak G, Rowland J, Harrison L, Hoell A (2015) The climate hazards infrared precipitation with stations—A new environmental record for monitoring extremes. Sci Data 2(1):1–21
Galdino S, Vieira LM, Pellegrin LA (2005) Impactos ambientais e socioeconômicos na Bacia do Rio Taquari-Pantanal. Corumbá: Embrapa Pantanal, 2005
Gregory KJ (2006) The human role in changing river channels. Geomorphology 79(3–4):172–191
Godoy JM, Padovani CR, Guimarães JR, Pereira JC, Vieira LM, Carvalho ZL, Galdino S (2002) Evaluation of the siltation of River Taquari, Pantanal, Brazil, through 210Pb geochronology of floodplain lake sediments. J Braz Chem Soc 13:71–77
Google Earth Engine (2019) A planetary-scale platform for earth science data & analysis. Disponível em. https://earthengine.google.com/
Hamilton SK, Sippel SJ, Calheiros DF, Melack JM (1997) An anoxic event and other biogeochemical effects of the Pantanal wetland on the Paraguay River. Limnol Oceanogr 42(2):257–272
Hu S, Niu Z, Chen Y, Li L, Zhang H (2017) Global wetlands: potential distribution, wetland loss, and status. Sci Total Environ 586:319–327
IBGE (2019) Províncias Estruturais. Disponível em. https://www.ibge.gov.br/geociencias/todos-os-produtos-geociencias.html
Ivory SJ, McGlue MM, Spera S, Silva A, Bergier I (2019) Vegetation, rainfall, and pulsing hydrology in the Pantanal, the world's largest tropical wetland. Environ Res Lett 14(12):124017
Keddy PA, Fraser LH, Solomeshch AI, Junk WJ, Campbell DR, Arroyo MT, Alho CJ (2009) Wet and wonderful: the world's largest wetlands are conservation priorities. Bioscience 59(1):39–51
Latrubesse EM, Amsler ML, de Morais RP, Aquino S (2009) The geomorphologic response of a large pristine alluvial river to tremendous deforestation in the South American tropics: the case of the Araguaia River. Geomorphology 113(3):239–252
Lehner B, Döll P (2004) Development and validation of a global database of lakes, reservoirs and wetlands. J Hydrol 296(1):1–22
Louzada RO, Bergier I, Assine ML (2020) Landscape changes in avulsive river systems: case study of Taquari River on Brazilian Pantanal wetlands. Sci Total Environ 723:138067

Louzada RO, Bergier I, Roque FO, McGlue MM, Silva A, Assine ML (2021) Avulsions drive ecosystem services and economic changes in the Brazilian Pantanal wetlands. Curr Res Environ Sustain 3:100057

Markert KN, Schmidt CM, Griffin RE, Flores AI, Poortinga A, Saah DS, Muench RE, Clinton NE, Chishtie F, Kityuttachai K, Someth P, Anderson ER, Aekakkararungroj A, Ganz DJ (2018) Historical and operational monitoring of surface sediments in the lower mekong basin using landsat and google earth engine cloud computing. Remote Sens 10(6):909

Mason J, Mohrig D (2019) Differential bank migration and the maintenance of channel width in meandering river bends. Geology 47(12):1136–1140

McGlue M, Silva A, Corradini F, Zani H, Trees M, Ellis G, Parolin M, Swarzenski P, Cohen A, Assine M (2011) Limnogeology in Brazil's "forgotten wilderness": a synthesis from the large floodplain lakes of the Pantanal. J Paleolimnol 46(2):273–289

McGlue MM, Silva A, Zani H, Corradini FA, Parolin M, Abel EJ, Cohen AS, Assine ML, Ellis GS, Trees MA, Kuerten S, Gradella FDS, Rasbold GG (2012) Lacustrine records of Holocene flood pulse dynamics in the upper Paraguay River watershed (Pantanal wetlands, Brazil). Quat Res 78(2):285–294

Makaske B, Maathuis BH, Padovani CR, Stolker C, Mosselman E, Jongman RH (2012) Upstream and downstream controls of recent avulsions on the Taquari megafan, Pantanal, south-western Brazil. Earth Surf Proc Land 37(12):1313–1326

MapBiomas (2021) Coleção 7.0 da Série Anual de Mapas de Uso e Cobertura da Terra do Brasil. https://mapbiomas.org/

Montanher OC, Novo EMLM, Barbosa CCF, Rennó CD, Silva TSF (2014) Empirical models for estimating the suspended sediment concentration in Amazonian white water rivers using Landsat 5/TM. Int J Appl Earth Obs Geoinf 29:67–77

Pupim FN, Assine ML, Sawakuchi AO (2017) Late quaternary Cuiabá megafan, Brazilian Pantanal: channel patterns and paleoenvironmental changes. Quat Int 438:108–125

Ramsar COW (2018) Global wetland outlook: state of the world's wetlands and their services to people. Ramsar Convention Secretariat, Gland, Switzerland

Risso A, Bordas MP, Borges AL (1997) Produção de sedimentos. BRASIL. Ministério do Meio Ambiente, dos Recursos Hídricos e da Amazônia Legal. Plano de Conservação da Bacia do Alto Paraguai (Pantanal)-PCBAP: hidrossedimentologia do Alto Paraguai. Brasília, DF: MMA/PNMA 2:281–317

Roque FO, Ochoa-Quintero J, Ribeiro DB, Sugai LS, Costa-Pereira R, Lourival R, Bino G (2016) Upland habitat loss as a threat to Pantanal wetlands. Conserv Biol 30(5):1131–1134

RStudio Team (2021) RStudio: Integrated Development for R. RStudio, PBC, Boston, MA URL Disponível em. http://www.rstudio.com/

Schumm SA (2007) River variability and complexity. Cambridge University Press, Cambridge

Silva JSV, Abdon MM, Galdino S (2005) Desmatamento na Bacia do Alto Taquari no período de 1976 a 2000. In: Impactos ambientais e socioeconômicos na Bacia do Rio Taquari-Pantanal, EMBRAPA Pantanal, Corumbá, pp 125–138

Souza RMD, Seibert D, Quesada HB, de Jesus Bassetti F, Fagundes-Klen MR, Bergamasco R (2020) Occurrence, impacts and general aspects of pesticides in surface water: a review. Process Saf Environ Prot 135:22–37

Suizu TM, Latrubesse EM, Stevaux JCN, Bayer M (2022) Resposta da morfologia do médio-curso superior do Rio Araguaia às mudanças no regime hidrossedimentar no período 2001–2018. Revista Brasileira de Geomorfologia 23(2):1420–1434

USGS (2020) Landsat 4–7 collection 1 (C1) surface reflectance (LEDAPS) product guide. America 3:1–33

Wantzen KM, Drago E, Da Silva CJ (2005) Aquatic habitats of the upper Paraguay River-floodplain-system and parts of the Pantanal (Brazil). Int J Ecohydrol Hydrobiol 5(2):107–126

Wohl E (2015) Legacy effects on sediments in river corridors. Earth Sci Rev 147:30–53

Zani H, Assine ML, McGlue MM (2012) Remote sensing analysis of depositional landforms in alluvial settings: method development and application to the Taquari megafan, Pantanal (Brazil). Geomorphology 161:82–92

Zhou L, Lau KM (1998) Does a monsoon climate exist over South America? J Clim 11(5):1020–1040

Chapter 5
Coastal Storm Events and Associated Impacts on Oceanic Beaches Between Ilha Grande and Cabo Frio (RJ, Brazil)

Breylla Campos Carvalho, Thalles Augusto Abreu Araújo, and Josefa Varela Guerra

Abstract Understanding the impacts of storms on coastal environments, particularly on sandy beaches, is highly relevant due to their vulnerability to erosion during such events. This research compared two distinct parameters, Storm Power Index (SPI) and Wave Energy Flux (WEF), to identify, quantify, and classify coastal storm events based on the behavior of gravity waves. The investigated region comprises all oceanic beaches between Ilha Grande and Cabo Frio, in the state of Rio de Janeiro, SE Brazil, spanning over 200 km of coastline. During the investigated period (1986–2018), over 400 storm events were identified, presenting a clear interannual variability, mostly marked by energetic events in *La Niña* years. The four most energetic episodes were studied in detail in terms of wave setup, near-bottom orbital velocity, and net shoreline movement. It was found that these episodes induced sediment remobilization between the coastline and the 30 m isobath, leading to retreat in more than 80% of the study area's coastline. Finally, the WEF parameter was found to be more consistent than the SPI for this kind of study as it physically represents the amount of energy transferred from the ocean waves to the coast.

Keywords Wave energy flux · Shoreline variability · Coastal erosion · Wave set-up · La Niña

B. C. Carvalho (✉)
Instituto do Mar, Universidade Federal de São Paulo (Unifesp), Rua Doutor Carvalho de Mendonça 144, 11070-100 Santos, São Paulo, Brazil
e-mail: breylla.carvalho@uerj.br

B. C. Carvalho · J. V. Guerra
Departamento de Oceanografia Geológica, Faculdade de Oceanografia, Universidade do Estado do Rio de Janeiro (UERJ), Rio de Janeiro, Rio de Janeiro, Brazil
e-mail: josefa@uerj.br

T. A. A. Araújo
School of Earth Sciences, University of Western Australia, Perth, WA, Australia

Forrest Research Foundation, Perth, WA, Australia

T. A. A. Araújo
e-mail: thalles.araujo@research.uwa.edu.au

A. Coronato and G. B. Alves (eds.), *Latin American Geomorphology*, The Latin American Studies Book Series, https://doi.org/10.1007/978-3-031-55178-9_5

5.1 Coastal Storm Events and Associated Coastal Impacts

Storms can be understood as the representation of one of the most energetic states of nature, commonly associated with strong winds, heavy rains, storm surges, and, when in the coastal ocean, erosion of beaches and flooding of the coastal zone (Coco and Ciavola 2017). According to Harley (2017), in order to understand the degree of exposure of a certain region to storm events, it is necessary to investigate the temporal patterns and tendencies related to these events. This includes but is not limited to the frequency of the events within specific periods of time, the time intervals between storms, seasonal patterns, teleconnections with climate patterns (e.g. *El Niño* and *La Niña*), and spatial patterns. According to Melet et al. (2018), waves are a leading cause of extreme sea level episodes and, over the recent decades, changes in wave regimes have been observed as a response to climate change.

During storm conditions, the amount of energy transferred from the ocean to the coastline increases and, consequently, the coastal zone might become vulnerable. Devastating scenarios can occur when the coastal topography is flat and the population density is high along the coast. As such, it is imperative to investigate the magnitude and the variability of coastal storms, as well as the consequent shoreline response. This is especially significant in times of climate change, during which scientific studies have worldwide confirmed an increased frequency of extreme events (Young and Ribal 2019), sea level rise (Losada et al. 2013), and increased coastal erosion (Mentaschi et al. 2018). The accurate prediction of coastal storms and their associated impacts, such as shoreline variability, is extremely challenging due to the intricate nature of processes that operate over a wide spectrum of temporal and spatial scales.

Computational models are one of the most used tools to assess the intensity of coastal storms and their impacts on the coastline. These models are software that code fundamental laws of physics in order to calculate how the spatial conditions evolve over discrete time steps. Environmental information such as wave height, period, and direction serve as inputs to the calculations, so forecasts of the coastal ocean behavior can be made. The accuracy of these models is paramount in foreseeing the impacts that natural hazards such as coastal storms impose over coastal communities, not only preventing loss of property but also protecting lives (Elko et al. 2019). However, it is essential to evaluate such systems considering the contributions of all spatial and temporal scales involved. Contributions from different sources might trigger key drivers. For example, variable spatial and temporal conditions might interact, leading to alongshore variability of incident waves and a consequent alongshore gradient in the shoreline response (Wiles et al. 2022). Additionally, long-term coastal resilience is a major variable to be considered since isolated events must be investigated within a wider framework encompassing climate change and sea level rise (Brooks et al. 2017).

5.1.1 Tools for Detecting Storm Events

5.1.1.1 Storm Power Index

The Storm Power Index (SPI) is a methodology first introduced by Dolan and Davis (1992) to investigate storm events based on the behavior of gravity waves. It is based on the temporal integration of the square of the significant wave height (H_s) (Eq. 5.1).

$$\text{SPI} = \int_{t_1}^{t_2} H_s^2 dt \tag{5.1}$$

It quantifies the amount of energy that is transferred through the waves from the ocean to the coastline. For example, if a wave system persists with $H_s = 3$ m for 12 h, its SPI is 3 m squared times 12 h which is equal to 216 m^2 h. The integral operation takes into account continuous variations of H_s integrated over the whole duration of a storm event. As such, it is a parameter that represents some sort of accumulated amount of energy, although it does not physically represent a certain amount of energy since it is not expressed in Joules.

5.1.1.2 Wave Energy Flux

The Wave Energy Flux (WEF), on the other hand, represents the amount of energy that is transferred through the waves from the ocean to the coastline. It is calculated as a function of the wave directional spectrum. When adapted to be used alongside meteo-oceanographic buoy datasets, it is in accordance with the Linear (Airy) Wave Theory and takes into account all parameters that describe a wave system: H_s, wave period (T), wavelength (λ), wave direction (Dir), and the local depth (d). Furthermore, it is valid for all water depths that are in accordance with the linear wave theory (the so-called "deep", "intermediate", and "shallow" waters), and it is expressed by Eq. 5.2:

$$\vec{P} = \frac{\rho g H_s^2}{16}\left[\frac{1}{2} + \frac{2\pi d}{\lambda \sinh\left(\frac{4\pi d}{\lambda}\right)}\right]\vec{c} \tag{5.2}$$

where ρ is the seawater density (~1025 kg/m^3), g is the gravitational acceleration (9.81 m/s^2) and $\vec{c}$ is the wave celerity, which incorporates the wave period and direction. As a vector parameter, it has a direction, which coincides with the direction of the wave propagation. Its dimension represents "energy per unit time per unit meter" (J/(m × s)). In the SI, these units represent "*the amount of Joules that pass, per second, through a plane of dimensions (1* m × *d* m*) that is perpendicular to the direction of propagation of the wave and where d is the local depth*" (Fig. 5.1). As the plane is perpendicular to the direction of propagation of the wave, the WEF can

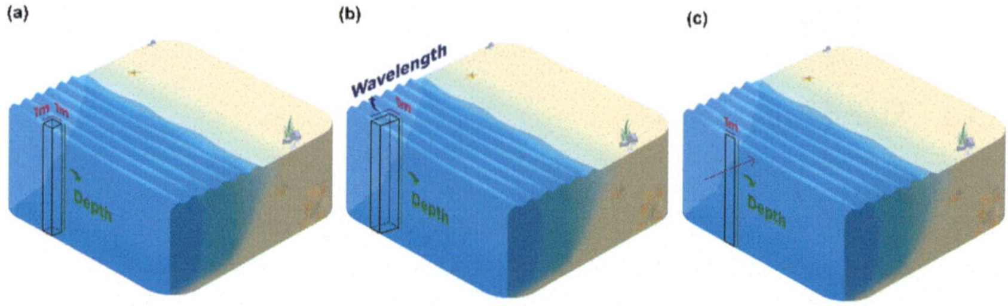

Fig. 5.1 Illustration of different parameters relative to the flux of wave energy (S.I. units assumed): **a** the wave energy density, usually denoted by E, represents the amount of energy (J) contained in a parallelepiped of sides [1 m × 1 m × d m] and therefore has units Jm^{-2} (it is not Jm^{-3} because the vertical direction has been integrated); **b** the wave energy density E multiplied by the wavelength λ (that is, the quantity $E\lambda$) represents the amount of energy (J) contained in a parallelepiped of sides [1 m × λ m × d m] and therefore has units Jm^{-1} (the vertical and one of the horizontal directions have been integrated); **c** the wave energy flux (WEF), usually denoted by $\overrightarrow{P}$, represents the amount of energy (J) that passes through a plane of area [1 m × d m] per unit time (s) and therefore has units $\text{Jm}^{-1}\ \text{s}^{-1}$ (the vertical direction has been integrated)

be interpreted as a measure of the energy that reaches the coast per unit of time per unit of length in the alongshore direction.

For the investigation of coastal storms reaching the coastline, however, only its component that is perpendicular to the coast must be considered. As such, this study considers (Eq. 5.3)

$$P_{\perp} = \vec{P} \cdot \hat{n} = P_x n_x + P_y n_y \tag{5.3}$$

where $\hat{n}$ is a unit vector that is perpendicular to the coast.

As such, the WEF is not a parameter that represents an *accumulated* energy. Instead, it is a parameter that represents an *instantaneous* rate of energy (that is, energy per unit time, per unit of distance in the alongshore direction) that reaches the coastline. In order for us to have a parameter that is comparable to the SPI, we need to transform the WEF into a parameter that measures the *accumulated* amount of energy that reaches a coastline. This is done by integrating Eq. 5.3 (Eq. 5.4)

$$\text{Accumulated WEF} = \int_{t_1}^{t_2} P_{\perp}\, dt = \int_{t_1}^{t_2} \vec{P} \cdot \hat{n}\, dt \tag{5.4}$$

So, Eq. 5.4 represents the amount of wave energy that is *accumulated* over time, and that effectively reaches the coastline during a storm event. Since it takes into account only the WEF component that is perpendicular to the coastline, Eq. 5.4 is comparable to Eq. 5.1.

5.2 Case Report

Worldwide, the last decades have been marked by great efforts in trying to understand and predict the impacts of extreme events on shoreline variability in different coastal environments (e.g. Honeycutt et al. 2001; Roelvink et al. 2009; Phantuwongraj et al. 2013; Puig et al. 2016; Splinter et al. 2017; Plant et al. 2017; Dietz et al. 2018; Garnier et al. 2018) including in Brazil (Calliari et al. 1998; Gentile and Goya 2005; Esteves et al. 2006; Parise et al. 2009; Machado et al. 2010; Gentil de Farias and Parente Maia 2010; Gregório et al. 2017; Paula and Farrapeira Neto 2017). The occurrence of coastal storm events and their effects on Rio de Janeiro's coastline have also been studied for particular sites (e.g. Carvalho et al. 2020; Fernandez et al. 2011; Lima et al. 2006; Muehe, 2011; Muehe et al. 2015; Pena and Lins-de-Barros 2015; Pereira et al. 2018; Santos et al. 2004), focusing on their vulnerability to these events, which are more frequent between April and September.

The present study was designed to compare the ability of the SPI and WEF parameters to detect storm events and to evaluate their possible impacts on Rio de Janeiro's Coastline. As such, the main goals of this work are to identify, quantify, classify, and characterize storm events that occurred between 1986 and 2018 between Ilha Grande and Cabo Frio (Fig. 5.2). More details about the study area and the methodology employed are presented in the next sections.

5.2.1 Study Area

The study area is located in the southern sector of Rio de Janeiro state, southeastern Brazil (Fig. 5.2a). According to Muehe and Valentini (1998), the oceanic beaches from this sector exhibit a west–east orientation with the development of sandy features that, except in Restinga da Marambaia barrier island (Dadalto et al. 2022), frequently occur as a pair of parallel sandy barriers separated by a depression, providing the rectified characteristic of its coastline. The main morphological and sedimentological characteristics of the beaches of this region are shown in Table 5.1.

The local climate is characterized by high temperatures and relative humidity, with calm winds from the northern quadrant and intense precipitation during the summer. The South Atlantic Subtropical Anticyclone (SASA) (Dereczynski and Menezes 2015) also influences this region throughout the year. The presence of frontal systems disturbs the stability of the SASA, often causing increased nebulosity, high precipitation, and strong winds.

Regarding the wave climate, the fair-weather scenario is characterized by short-period waves from northeast and east, whereas storm scenarios are characterized by long-period waves from the south (Parente et al. 2015; Carvalho et al. 2020). In the study area, wave height varies from 0.5 to 2.3 m, wave period ranges between 9 and 17 s, and wave direction oscillates between 120° and 190° (Criado-Sudau et al. 2019;

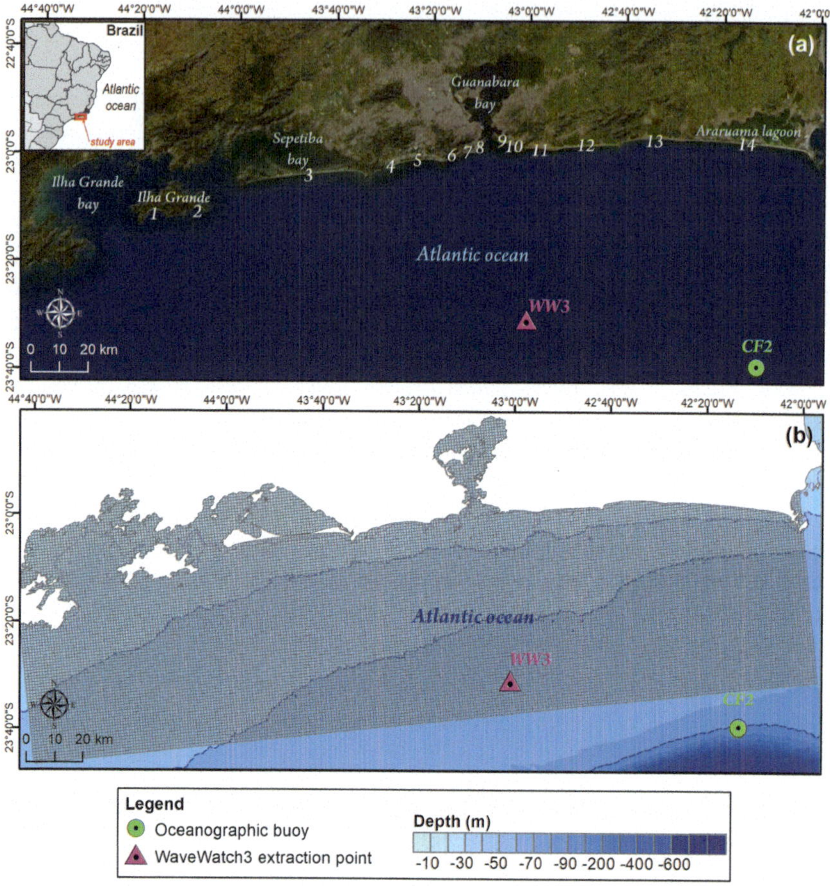

Fig. 5.2 Coastline of the study area: **a** TerraColor satellite imagery from Earthstar Geographics, available in ArcMap™ Basemap, showing the oceanic sandy beaches: 1. Praia do Leste and Praia do Sul; 2. Lopes Mendes; 3. Restinga da Marambaia; 4. Macumba; 5. Recreio-Barra da Tijuca; 6. São Conrado; 7. Leblon-Ipanema-Arpoador; 8. Copacabana-Leme; 9. Piratininga; 10. Camboinhas-Itaipu; 11. Itaipuaçú; 12. Barra de Maricá; 13. Saquarema; 14. Massambaba; **b** regional grid of the numerical model applied in this study. Local depth at point WW3 = 116 m, and at CF2 = 200 m

Nemes et al. 2019; Carvalho et al. 2021). The region is under a microtidal regime with an east–west direction of propagation (Harari and Camargo 1994).

Table 5.1 Morphological and sedimentological characteristics of the oceanic beaches between Ilha Grande and Cabo Frio (Fig. 5.1)

Beach	Extension (km)	Δ D_{50} (mm)	Average orientation	Δ Slope (°)	Δ Width (m)	Source
Praia do Leste and Praia do Sul	5.8	0.3–0.6	SE–NW	3.0–8.0	20–30	da Silva Amador (1987)
Lopes Mendes	2.5	0.1–0.5	SE–NW	1.8–2.6	37–80	Gralato (2016)
Restinga da Marambaia	40.0	0.3–0.6	W–E	n.d.a	50–140	Dadalto et al. (2021)
Macumba	2.5	0.5–0.7	W–E	1.9–10.6	23–110	Carvalho et al. (2021)
Recreio-Barra da Tijuca	18.0	0.3–0.6	W–E	2–10.1	29–120	Carvalho et al. (2021)
São Conrado	2.2	0.3–0.4	W–E	3–12.3	23–102	Carvalho et al. (2021)
Leblon-Ipanema-Arpoador	3.7	0.2–0.4	W–E	1.9–22	14–128	Carvalho et al. (2021)
Copacabana-Leme	3.8	0.3–0.4	SW–NE	0.4–10.6	9–150	Carvalho et al. (2021)
Piratininga	2.9	n.d.a	SE–NW	2.3–4.2	34–87	Rodrigues et al. (2015), Eccard et al. (2017)
Camboinhas-Itaipu	3.1	0.1–0.5	SE–NW	2.3–7.1	24–75	Rodrigues et al. (2015), Eccard et al. (2017)
Itaipuaçu	11.5	0.5–4.0	W–E	14–40	35–103	Silva et al. (2008)

(continued)

Table 5.1 (continued)

Beach	Extension (km)	Δ D_{50} (mm)	Average orientation	Δ Slope (°)	Δ Width (m)	Source
Barra de Maricá	21.5	0.3–0.7	W–E	3.7–4	66–90	Carvalho et al. (2014), Batista et al. (2015)
Saquarema	19.5	0.3–0.7	W–E	4–6	40–95	Pinto et al. (2015)
Massambaba	48.1	0.3–0.5	W–E	1.8–5.6	82–196	Muehe and Corrêa (1989), Maluf et al. (2016)

Legend Δ: range; n.d.a.: no data available; D_{50}: median sediment size

5.2.2 *Materials and Methods*

5.2.2.1 Wave Climate

Information on significant wave height (H_s), peak period (T_p), and peak direction (D_p), from January 1986 to December 2018, were extracted from the model WAVE-WATCH III (WW3), freely provided by NOAA, and validated with a time series registered from 2016 to 2018 by the CF2 buoy (PNBOIA program managed by the Brazilian Navy) (Fig. 5.2b). Additionally, the magnitude of the wave energy flux ($\vec{P}$) was calculated via Eq. 5.2, using the code developed by Araujo and Carvalho (2022).

5.2.2.2 Detection of Storm Events

The identification of storm events was conducted with statistical parameters extracted from the WW3 time series. The thresholds to define each event were obtained from the 95° percentile of H_s (≥3 m) and $\vec{P}$ (≥48 kW/m). Then, filters were applied to detect which events could be classified as storms. In order to be considered so, key parameters must be maintained above certain thresholds for 12 h or more. The magnitude of each event was estimated by two parameters: the SPI (Eq. 5.1) and the Accumulated WEF (Eq. 5.4).

Table 5.2 Classes of magnitude of the storm events

Classes	SPI (m^2h)	Accumulated WEF (kJ/m)
I—Weak	<280	<570
II—Moderate	280–590	570–1215
III—Significant	590–910	1215–1965
IV—Severe	910–1190	1965–3125
V—Extreme	>1190	>3125

A Student's *t*-test was then applied to the results to evaluate the significance of the differences between the values obtained by each method. Likewise, Pearson's coefficients of correlation (r) were calculated. Finally, a cluster analysis was conducted from the Euclidean distance to classify the storm events into five distinct classes (Table 5.2).

5.2.2.3 Climate Index

The Southern Oscillation Index (SOI) from 1986 to 2018, provided by the National Oceanic and Atmospheric Administration (NOAA), was used to identify the occurrence of *El Niño* and *La Niña* years.

5.2.2.4 Propagation of Waves

The possible impacts of storm events on the coastline and on its adjacent continental shelf were evaluated considering the wave orbital velocity close to the bottom during the four most energetic events. For this, the module WAVE of the software Delft3D was applied for a rectangular mesh with the following spatial scales: regional (1 km × 1 km), intermediate (0.25 km × 0.25 km), and local (25 m × 25 m). The bathymetry used was obtained from nautical charts provided by the Brazilian Navy and compiled by the GEOMARGEM research group (Fig. 5.2b). The bathymetric survey over the continental shelf (beyond the 50 m isobath) was conducted between 1973 and 1979. The bathymetric survey between the 50 m isobath and the shoreline was conducted between 1980 and 1998. The bathymetric surveys within Sepetiba and Guanabara coastal bays were conducted between 2004 and 2015. The model's boundary conditions are presented in Table 5.3 and the physical settings are presented in Table 5.4.

5.2.2.5 Impacts on the Coastline

The coastline position before and after each selected storm event (Table 5.3) was mapped using Landsat 5 satellite images, path/row 217/076. The dry–wet interface, used as the shoreline indicator (Boak and Turner 2005), was obtained by applying

Table 5.3 Boundary conditions for each modeled scenario

Scenario	H_s (m)	T_p (s)	D_p (°)
1988	3.9	12.8	161
1997	4.2	12.1	144
2001	4.0	11.0	156
2011	3.7	13.8	164

Table 5.4 Model settings

Parameter	Value
Gravitational acceleration	9.81 m/s^2
Sea water density	1025 kg/m^3
Minimum depth	0.05 m
Spectral resolution	36 directions
Wind	Disabled
Bottom friction	0.067 m^2/s^3 (JONSWAP)
Diffraction coefficient	0.2
Refraction	Enabled
Bottom-induced wave break	Battjes and Janssen (1978)
Whitecapping formula	Komen et al. (1984)

the Normalized Difference Water Index (NDWI) that identifies the interface between water and land (McFeeters 1996) (Eq. 5.5):

$$\text{NDWI} = \frac{(\text{greenband} - \text{NIR})}{(\text{greenband} + \text{NIR})} \tag{5.5}$$

where NIR corresponds to the near-infrared band. The NDWI was calculated with the ArcMap™ 10.5 Raster Calculator tool, resulting in raster outputs that were converted into vectors. These vectors were used to calculate the Net Shoreline Movement (NSM) for each event, determined through the tool Digital Shoreline Analysis System 4.0 (DSAS), an ArcMap™ extension developed by the United States Geological Survey (USGS) (Himmelstoss et al. 2021). A total of 1360 cross-shore coastline transects, equi-spaced 100 and 300 m long were used. The dates of the used images are shown in Table 5.5. The NSM values are as follows:

- Erosion: coastline variation of <–5 m (more negative);
- Stability: coastline variations between –5 and +5 m;
- Accretion: coastline variations of >+5 m (more positive).

Table 5.5 Dates of the Landsat 5 images used to calculate the NSM for the four most energetic events listed in Table 5.3

Date before the storm	Date after the storm	Interval between images (days)
28/Jul/1988	10/Apr/1989	256
11/Feb/1997	03/Jun/1997	112
27/Apr/2001	29/May/2001	32
13/Aug/2011	30/Sep/2011	48

5.2.3 *Results*

5.2.3.1 Wave Climate

Throughout the studied period (1986–2018) the average significant wave height was ~1.7 m, oscillating between 0.4 and 5.0 m. Sixty-four percent of the wave heights ranged between 1.0 and 2.0 m, and 23% between 2.0 and 3.0 m. The average peak period was ~9.8 s, varying between 3.1 and 18.0 s, while the most frequent periods oscillated between 6 and 9 s (38%). The average peak direction was SE (151°), whereas the most frequent directions were E (25%) and S (45%) (Fig. 5.3a, b).

5.2.3.2 Storm Events

Between 1986 and 2018, the storm events discriminated by $H_s \geq 3$ m and $\vec{P} \geq$ 48 kW/m were associated with waves from SSW, with periods of ~10 s, with a few occurrences of up to 18 s (Fig. 5.3c–f). The total number of events and their average duration, interval between events, and intensity, calculated with the two methods (SPI and WEF), are presented in Table 5.6. For both methods, one can note the occurrence of weak and moderate storm events throughout the year. However, the highest concentration of events of all classes was recorded between April and September when the highest accumulated magnitudes were reached (Fig. 5.4).

For the SPI method, a total of 167 events were observed in *El Niño* years and a total of 135 in *La Niña* years, with average magnitudes of 322 m^2h and 394 m^2h, respectively (Fig. 5.5a). For the WEF method, 158 events were observed during *El Niño* years and 143 during *La Niña* years, with average accumulated magnitude of 749 kJ/m and 719 kJ/m, respectively (Fig. 5.5b).

For the SPI method, the average duration of events was 28.8 h with a standard deviation (σ) of 18.6 h, while, for the WEF method, the average duration of the events was 28.1 h ($\sigma = 18.1$ h). In general, events with the longest duration occurred between May and July (~1930 h), whereas the shortest events occurred between December and February (~150 h) (Fig. 5.6a, c). The years with the longest cumulative duration were 2006 (SPI—579 h, WEF—618 h), 2007 (SPI—537 h, WEF—543 h), and 2010 (SPI—723 h, WEF—597 h) (Fig. 5.6). The comparison between the SPI and the WEF

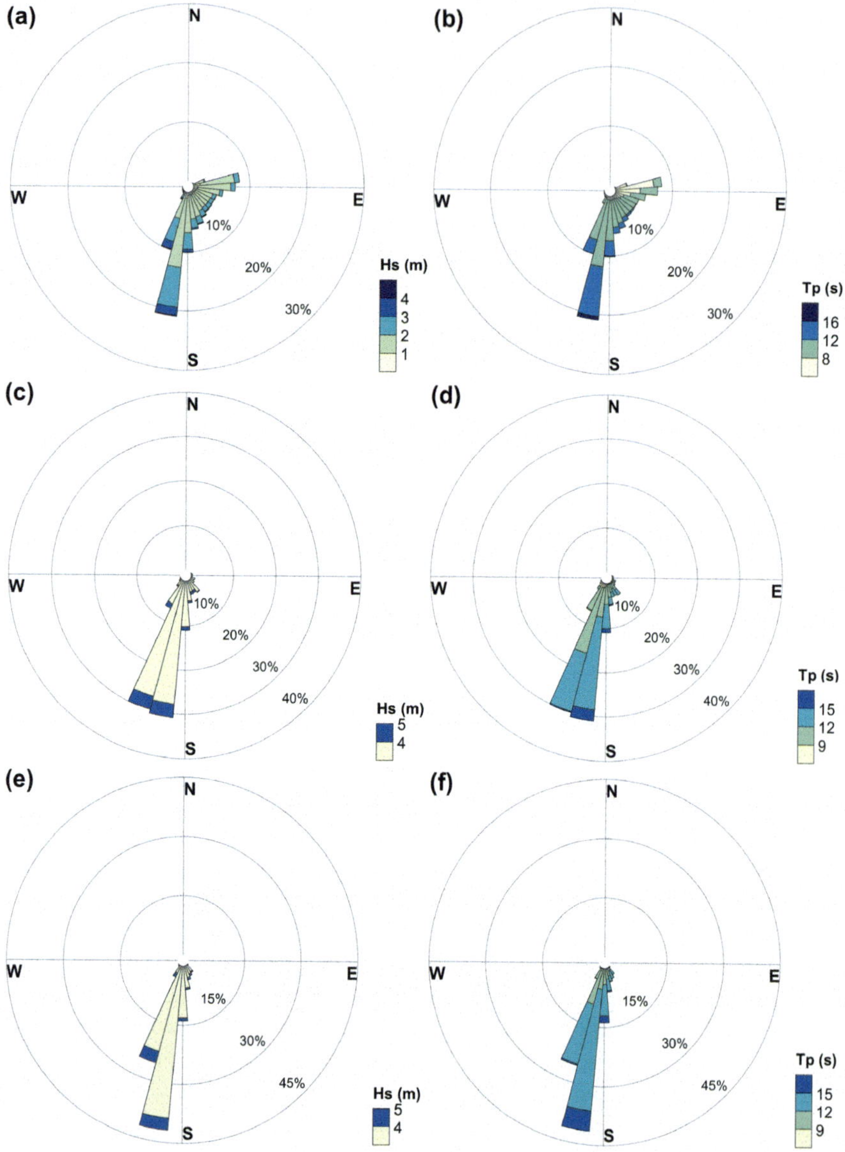

Fig. 5.3 Wave roses: (**a**) and (**b**) H_s and T_p for the whole time series (1986–2018); (**c**) and (**d**) H_s and T_p of storm waves filtered by the SPI; (**e**) and (**f**) H_s and T_p of storm waves filtered by the WEF parameter

Table 5.6 Occurrence and characteristics of the storm events according to the SPI and WEF methods for the period between 1986 and 2018

Method	Total number of events	Average duration of each event (hours)	Average interval between events (days)	Average intensity
SPI	425	28.7	19.9	351.6 (m^2h)
WEF	436	27.8	19.5	713.3 (kJ/m)

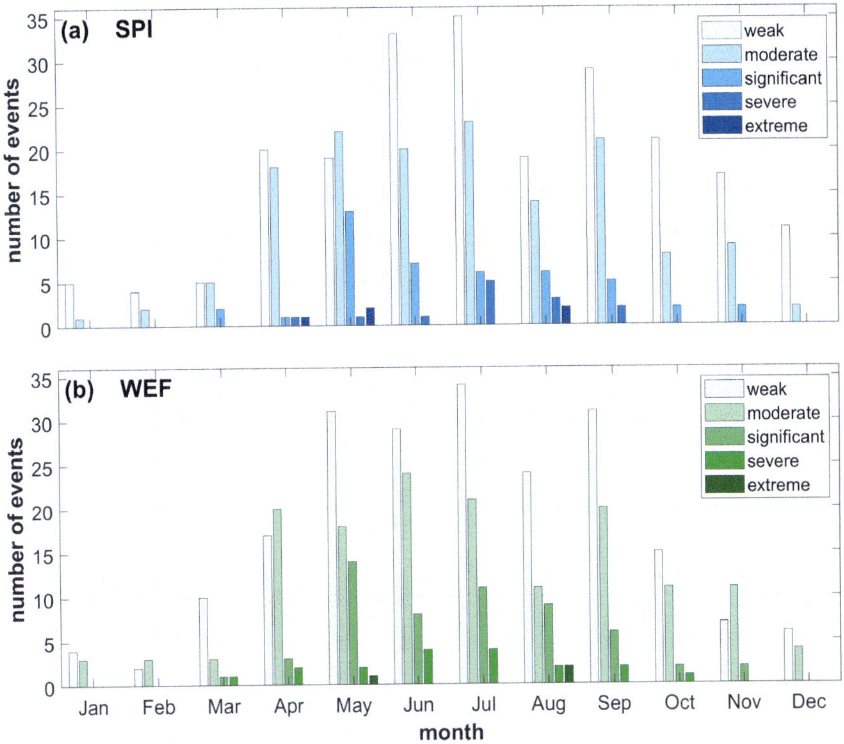

Fig. 5.4 Monthly distribution of storm events (1986–2018): **a** SPI; **b** WEF

methods, when determining the duration of each event, did not indicate significant variation (t-test > 0.5), presenting moderate correlation ($r = 0.75$).

As of the time interval between each event, the average value was 19.8 days ($\sigma =$ 4.2 days) from the filtering of the SPI data, and 19.2 days ($\sigma = 5.3$ days) for the WEF data. The years that presented the longest average intervals were 1993 (25–30 days) and 2005 (~25 days), whereas the shortest intervals occurred in 1996 (<10 days) and 2001 (~10 days) (Fig. 5.7). The comparison between the SPI and WEF results, when

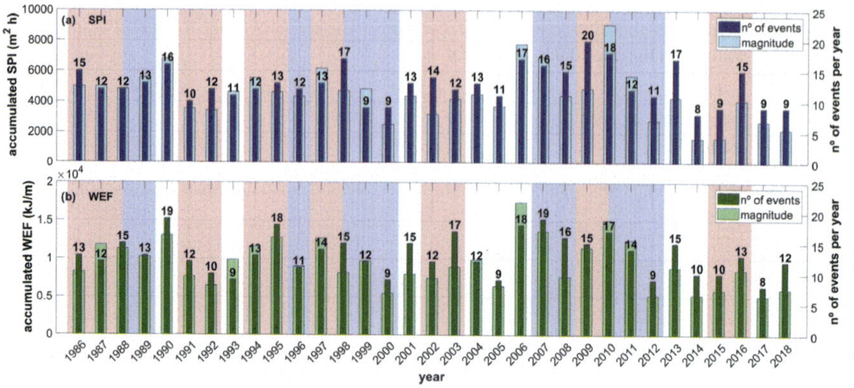

Fig. 5.5 Annual distribution of storm events (occurrence and accumulated magnitude): **a** SPI; **b** WEF. Red and blue shadows correspond, respectively, to *El Niño* and *La Niña* years

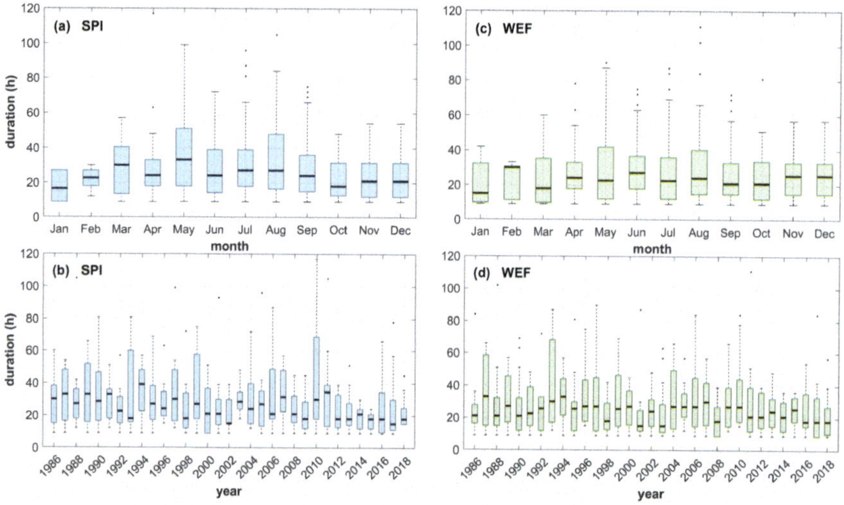

Fig. 5.6 Mean duration of storm events: **a** monthly and **b** annual values based on the SPI; **c** monthly and **d** annual values based on the WEF parameter

determining the average intervals between events, also did not indicate a significant difference (t-test > 0.6), presenting a moderate correlation ($r = 0.66$).

We verified that one method does not always predict that a certain event will be classified in the same class as the other method, presenting moderate correlation ($r = 0.79$). The SPI method classified five events as extreme whereas the WEF method found only three (Fig. 5.8). Besides, the months that correspond to the greatest accumulated magnitudes according to the SPI do not match those determined by the WEF parameter. For this dataset, there was a significant variation ($p < 0.001$) and no correlation ($r = 0.009$). In contrast, the detection of the months with the highest

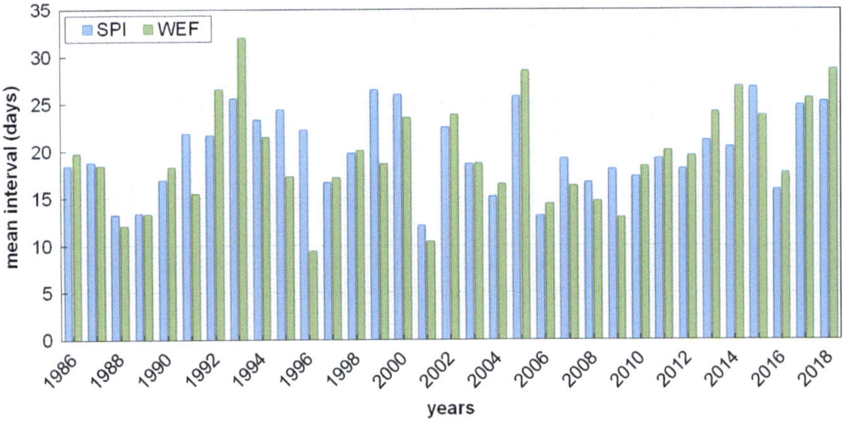

Fig. 5.7 Mean interval (days) between storm events

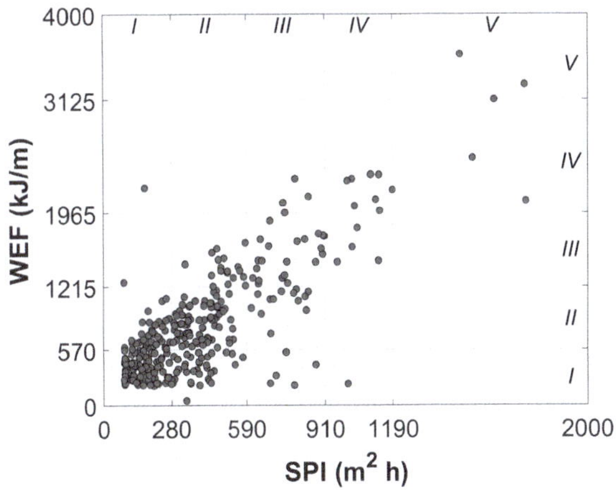

Fig. 5.8 Comparison between SPI and WEF values. Event classes (I–V) are shown in Table 5.2

number of events was similar (85% of the cases), presenting moderate correlation ($r = 0.74$).

5.2.3.3 The Four Most Energetic Events and Their Impacts on the Coastline

To assess the impacts of storm events on the shoreline, the four major events identified in the time series (Table 5.7) were selected for this purpose and will be chronologically presented in the following subsections.

Table 5.7 Oceanographic conditions during the four most energetic events between 1986 and 2018 and their ranking according to the SPI and WEF parameter

Date (dd/mm/yyyy)	Mean H_s (m)	Mean T_p (s)	Mean D_p (°)	Max D_p (°)	Duration (h)	SPI (m^2h)	WEF (kJ/m)
10/08 to 14/08/1988	3.9	12.8	161	205	105 (SPI) 102 (WEF)	1612 (2nd)	3126 (3rd)
30/05 to 03/06/1997	4.2	12.1	144	199	99 (SPI) 90 (WEF)	1738 (1st)	3279 (2nd)
05/05 to 09/05/2001	4.0	11.0	156	216	93 (SPI) 87 (WEF)	1522 (3rd)	2529 (4th)
21/08 to 26/08/2011	3.7	13.8	164	205	105 (SPI) 111 (WEF)	1468 (4th)	3589 (1st)

SPI: Storm Power Index; WEF: Wave Energy Flux

Oceanographic Conditions

The storm event that occurred between August 10 and 14, 1988, was classified as the second and the third most energetic event of the time series, respectively, for the SPI and WEF parameters (Fig. 5.9, Table 5.7). During this event, H_s varied from 1.9 to 4.9 m, T_p from 10.9 to 13.5, and D_p, from 137° to 205°. The calculated SPI was 1612 m^2h and the accumulated WEF was 3126 kJ/m, both configuring an extreme class of magnitude (Table 5.2). Santos et al. (2004) reported this event when waves up to 4 m high reached the Rio de Janeiro Coastline and destroyed harbors and urban structures.

The second selected storm event, that occurred between May 30 and June 3, 1997, was classified as the first and the second most energetic event of the time series, respectively, for the SPI and WEF parameters (Fig. 5.10, Table 5.7). During this event, H_s varied from 3.1 to 4.8 m, T_p, from 7.8 to 13.3 s, and D_p, from 144° to 227°. The calculated SPI was 1738 m^2h and the WEF was 3279 kJ/m, both configuring an extreme class of magnitude. Santos et al. (2004) also reported this event and the damage it inflicted to urban structures in Guanabara Bay and along the Rio de Janeiro Coastline.

Between May 5 and 9, 2001, it occurred the third selected storm event that was classified, according to the SPI and WEF parameters, respectively, as the third and the fourth most energetic event of the time series (Fig. 5.11, Table 5.7). During this event, H_s varied from 1.8 to 5.0 m, T_p between 10.2 and 11.7 s, and D_p, from 152° to 216°. The calculated SPI was 1522 m^2h (extreme event; Table 5.2) while the WEF was 2529 kJ/m, configuring a severe event (Table 5.2). Although this is not the strongest event in the series, it is the most reported in the literature (Innocentini et al. 2003; Muehe 2011; Palmeira et al. 2012; Carvalho et al. 2020).

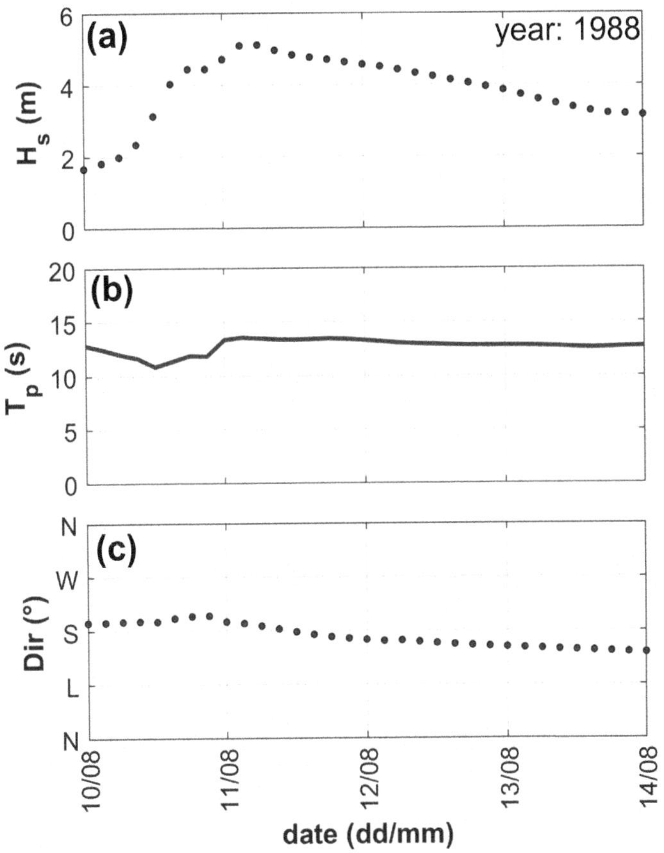

Fig. 5.9 Oceanographic conditions during the storm event of August 10–14, 1988: **a** H_s; **b** T_p; **c** D_p. Data from the WW3 extraction point (location in Fig. 5.1)

Last but not least, the fourth selected event occurred between August 21 and 26, 2011, and according to the SPI and WEF parameters, it was classified, respectively, as the fourth and the top-ranked event of the time series (Fig. 5.12, Table 5.7). During this event, H_s varied from 2.0 to 4.2 m, T_p, from 10.0 to 17.4 s, and D_p, from 169° to 205°. The calculated SPI was 1468 m^2h and the WEF was 3589 kJ/m, both configuring an extreme class of magnitude (Table 5.2).

Near-Bottom Orbital Velocity

During the four strongest events detected by both methods (Table 5.3), the near-bottom orbital velocities ranged between 0.5 and 0.8 m/s between 20 and 50 m deep, whereas close to the coastline the values remained above 0.8 m/s (Fig. 5.13). The storm events that occurred in 1997 and 2011, classified as the two stronger events

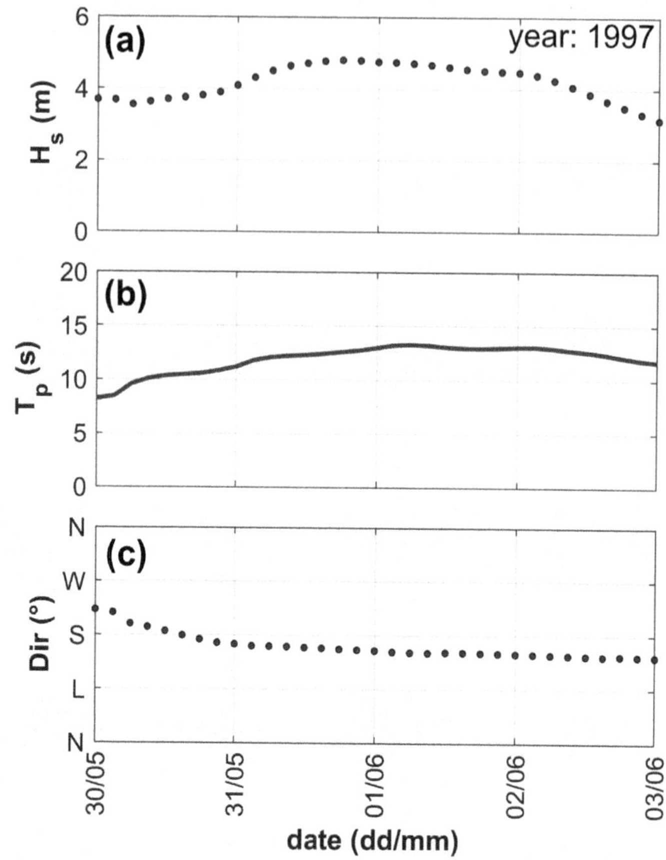

Fig. 5.10 Oceanographic conditions during the storm event of May 30–June 3, 1997: **a** H_s; **b** T_p; and **c** D_p. Data from the WW3 extraction point (location in Fig. 5.1)

according to the WEF parameter, caused higher velocities along the continental shelf when compared to the other two events (1988 and 2001).

Net Shoreline Movement

For the August 1988 event, 13% of the coastline was eroded, 24% was stable and 63% accreted (Fig. 5.14a). The most impacted beaches were Praia do Leste, Praia do Sul, and Restinga da Marambaia, with NSM between 5 and 25 m. The eastern sector of the study area (east from Guanabara Bay's entrance) was not mapped due to the high cloud coverage of the satellite image in that area. For the May and June 1997 events, 96% of the shoreline was eroded, with NSM above 25 m on most beaches (Fig. 5.14b). For May 2001, 73% of the shoreline was eroded, 22% remained stable and 5% accreted (Fig. 5.14c). The most affected beaches were Restinga da

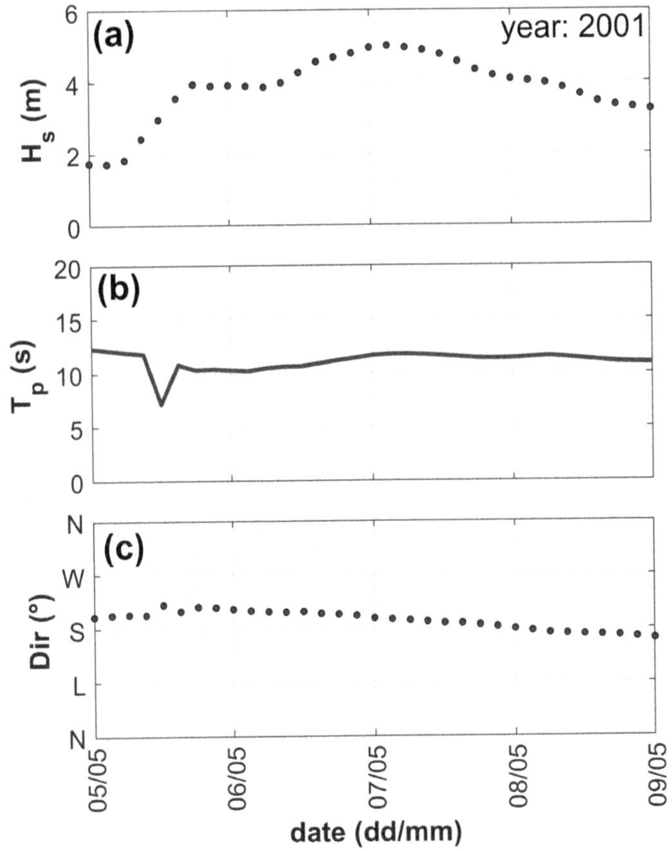

Fig. 5.11 Oceanographic conditions during the storm event of May 05–09, 2001: **a** H_s; **b** T_p; and **c** D_p. Data from the WW3 extraction point (location in Fig. 5.1)

Marambaia (central and eastern sectors), Recreio-Barra da Tijuca (central and eastern sectors), Saquarema (eastern sector), and Massambaba (eastern sector). Finally, for the August 2011 event, 32% of the shoreline was eroded, 35% remained stable and 32% accreted (Fig. 5.14d). The beaches located west of Guanabara Bay were more negatively affected (shoreline retreat between 5 and 25 m), when compared to those located to the east of the bay's entrance (shoreline accretion between 5 and 17 m). The events of 1997 and 2001 were the most harmful, causing coastline retreat along 80% of the studied oceanic beaches (Fig. 5.14b, c), reaching maximum values of 207 and 114 m, respectively.

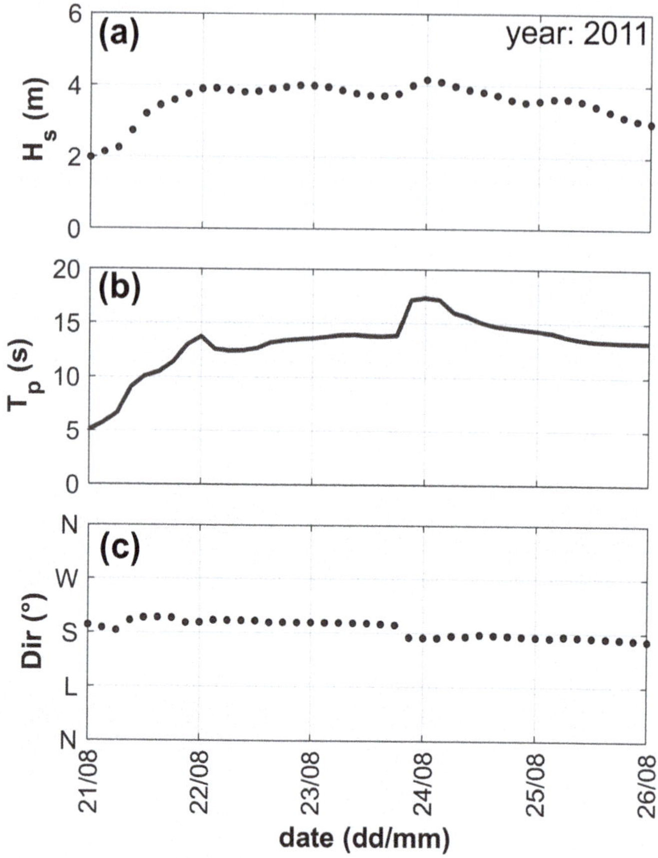

Fig. 5.12 Oceanographic conditions during the storm event of August 21–26, 2011: **a** H_s; **b** T_p; **c** D_p. Data from the WW3 extraction point (location in Fig. 5.1)

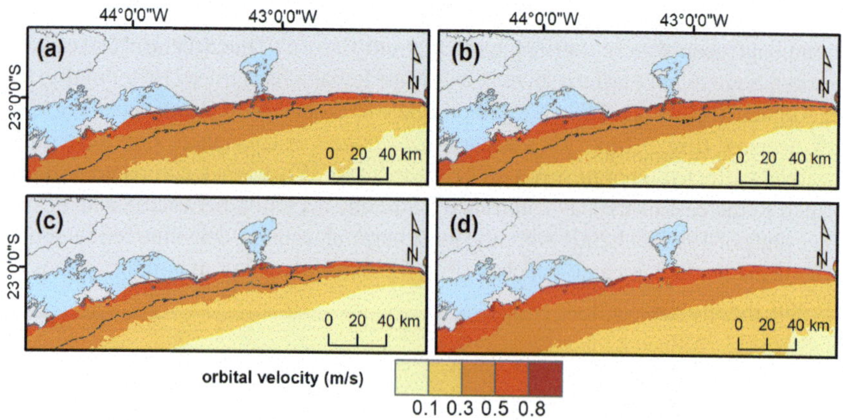

Fig. 5.13 Nearbed orbital velocities for each storm event: **a** 1988; **b** 1997; **c** 2001; **d** 2011

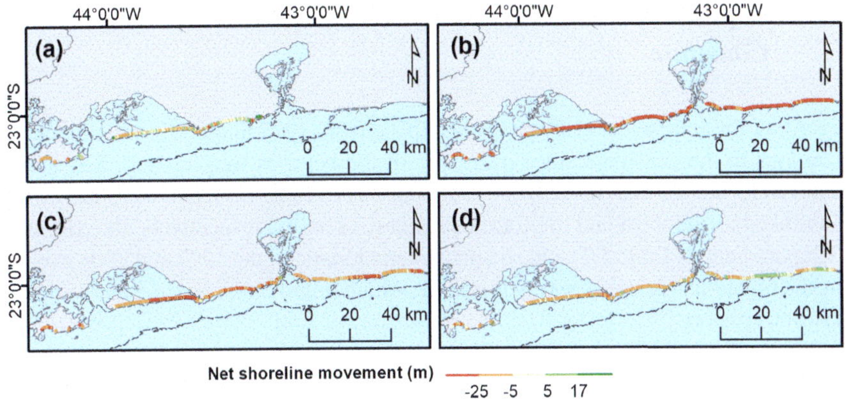

Fig. 5.14 Net shoreline movement for each storm event: **a** 1988—east from Guanabara Bay's entrance the satellite image was cloudy and it was not possible to detect the shoreline; **b** 1997; **c** 2001; **d** 2011

5.3 Discussion

5.3.1 SPI Versus WEF

In general, even though the two methods (SPI and WEF) are different, both can be satisfactorily used to describe storm events in terms of intensity, duration, and recurrence. The most notable difference between their results was the definition of the classes of the storm events, which is precisely dependent on the magnitude of the events. While the SPI considers only the square of the significant wave height, the WEF considers all wave parameters (height, period, wavelength, and direction) and the local depth that are necessary to calculate the physical energy associated with a wave system.

Additionally, it was observed that when applying the SPI method, the months associated with a high number of events do not necessarily correspond to those with the strongest events. For example, although four storm events were registered for July 1987 (SPI = 246 m^2h), the month with the greatest accumulated energy was March 1987, when only one event was registered (SPI = 788 m^2h). In the work by Mendoza et al. (2011) to verify the intensity of storm waves on the Mediterranean coast via the SPI method, the authors found no direct correspondence between the intensity of the cyclones that hit the coast and the classes of the storms. This result suggests that other parameters should be incorporated into the analysis. In this sense, the use of the WEF method is a prominent alternative to access the real magnitude of storm events since it takes into consideration not only significant wave height and duration of events but also wave period, then better representing the total amount of energy transferred from the ocean to the coast through gravity waves.

5.3.2 *Impacts of Extreme Events Along Rio De Janeiro's Coastline*

Regarding the four selected storm events, those from 1997 and 2001 were the most hazardous to the coastline, since over 70% of the beaches were eroded. Yet, these events were shorter than those from 1988 and 2011 (Table 5.7), and the 2001 event was ranked as the third and the fourth strongest event by, respectively, the SPI and WEF parameters (Table 5.7). One distinct characteristic of the 1997 and 2001 events was the higher incidence of waves from SW, which might have caused a more severe coastline erosion.

During the studied period (1986–2018) the *El Niño* events from 1997 to 1998 and 2015 to 2016 were two of the three most significant on record. Studies carried out in the USA's west coast found increased coastline erosion, particularly in the latter event, as a result of changes in the wave conditions (Barnard et al. 2015; Gordon et al. 2017). However, in the present work, it was verified that the *La Niña* of 2010–2012 was the most intense, with 38 storm events and a cumulative magnitude of 16766 m^2h for the SPI and 30919 kJ/m for the WEF. During this period, one of the most energetic events of the series was recorded (Aug/2011; Fig. 5.12; Table 5.7). However, this event was not one of the most damaging for the shoreline given that only ~32% of the coastline was eroded. This suggests that events should not be analyzed separately. Instead, it is necessary to understand the potential cumulative effects from the past, which are the recovery timescales of distinct beach profiles, how climate teleconnections and their combined effects influence storm events, and so on.

The spatial distribution of the estimated nearbed orbital velocities during the four selected storm events (Fig. 5.13) shows that, throughout the continental shelf shoreward of the 50 m isobath, threshold conditions for sand resuspension and transport (Komar and Miller 1975) were exceeded. Consequently, sediment became available in the water column to be horizontally transported, both alongshore and cross-shore, including during the downwelling-favorable conditions during which beach sands are carried to the nearby shoreface and continental shelf. In a study conducted by Yokoyama (2012) on the continental shelf of São Paulo, it was verified remobilization of very fine sands down to 22 m depth in conditions of waves from SE, whereas, in more energetic conditions (waves from S), there was remobilization down to the 60 m isobath. Guerra et al. (2006) and Storlazzi and Reid (2010) evaluated the remobilization of shelf sediments off the state of California (western US coast), and verified that it is associated with ENSO oscillations. While Guerra et al. (2006) concluded that the La Niña winter of 1998–1999 was more favorable for off-shelf sediment transport than the El Niño winter of 1997–1998, Storlazzi and Reid (2010) verified greater remobilization of fine sediments down to the 120 m isobath during periods of *El Niño*. These examples illustrate the importance of integrating the study of processes occurring along the coastline with those operating on the shoreface and the continental shelf.

Regarding the mobility of the coastline, Muehe and Klumb-Oliveira (2014) mention the resilience of the beaches between Restinga da Marambaia and Cabo Frio. They found that, even though *(1)* these beaches are exposed to extreme events, *(2)* they currently lack sources of continental sediment, and *(3)* their coastline retreated tens of meters (e.g. during the May 2001 storm event), they were able to recover without changing their foreshore position relative to sea level.

5.4 Conclusions

The SPI method takes into consideration H_s^2 and the duration of the storm (dimension of area times time), not necessarily physically representing the quantity of energy stored in the wave system. On the other hand, the WEF method takes into consideration all of the variables that describe a wave system (wave height, period, length, direction, and local depth of measurement) as well as the storm duration (dimension of energy per unit meter of wave crest), physically representing the energy transferred through a wave system from the ocean to the coastline. Therefore, the second method (WEF) presented itself as more effective, suitable, and reliable for calculating the magnitude of storm events that affected the Rio de Janeiro Coastline between 1986 and 2018.

Mapping the effects of storm events along the coastal zone is a task that usually faces hurdles such as the fact that satellite imagery is often unavailable for the specific dates of the events of interest. Additionally, numerical models are often based on the reanalysis of datasets and/or on nautical charts that are usually not up to date. Nonetheless, investigating such events and processes even with these limitations is still relevant when monitoring long coastlines. This is the reality of Brazil, where the financial and human resources are scarce for in situ monitoring and surveying. In this sense, it is worth making an effort to enhance image processing by, for example, developing and applying computational techniques based on artificial intelligence and machine learning. This way, demands of knowledge on coastal processes can be fulfilled to a greater extent and resources can be focused on strategic points in terms of coastline monitoring. This "smart" combination of remote sensing and in situ monitoring ends up being beneficial not only to researchers but also to the general society and to managers, who usually are end-users of such studies.

Acknowledgements This study was partially funded by the Coordenação de Aperfeiçoamento de Pessoal de Nível Superior—Brasil (CAPES)—Finance Code 001, as PhD scholarship (DS) and Doctoral Exchange Program PDSE [process #88881.132411/206-01, 19/2016] to the first author. We also thank Fundação de Amparo à Pesquisa do Estado do Rio de Janeiro (FAPERJ) for the project financial support [E-26/010.002208/2019]. The authors also thank Lucas de Martin Fortunato for the preparation of Fig. 5.1 and Leonardo Klumb de Oliveira (UFRB) for sharing the WW3 model wave records.

References

Araujo TAA, Carvalho BC (2024) A code to calculate the energy flux of ocean waves from oceanographic buoy data. Manuscript in preparation

Barnard PL, Short AD, Harley MD et al (2015) Coastal vulnerability across the Pacific dominated by El Niño/Southern Oscillation. Nat Geosci 8:801–807. https://doi.org/10.1038/ngeo2539

Batista AS, da Silva ALC, Pinto VCS et al (2015) Análise morfoscópica dos sedimentos de ambientes deposicionais no litoral de Maricá, estado do Rio de Janeiro. In: 14° Simpósio de Geologia do Sudeste. SBG, Campos do Jordão, pp 1–5

Battjes JA, Janssen JPFM (1978) Energy loss and set-up due to breaking random waves. 16th conference on coastal engineering. ASCE, Hamburg, pp 569–587

Boak EH, Turner IL (2005) Shoreline definition and detection: a review. J Coast Res 214:688–703. https://doi.org/10.2112/03-0071.1

Brooks SM, Spencer T, Christie EK (2017) Storm impacts and shoreline recovery: mechanisms and controls in the southern North Sea. Geomorphology 283:48–60. https://doi.org/10.1016/j.geomorph.2017.01.007

Calliari LJ, Tozzi HAM, Klein HF (1998) Beach morphology and coastline erosion associated with storm surges in Southern Brazil—Rio Grande to Chuí, RS. An Acad Bras Cienc 70:230–247

Carvalho BC, Dalbosco ALP, Guerra JV (2020) Shoreline position change and the relationship to annual and interannual meteo-oceanographic conditions in Southeastern Brazil. Estuar Coast Shelf Sci 235:106582. https://doi.org/10.1016/j.ecss.2020.106582

Carvalho BC, Lins-de-Barros FM, da Silva PL et al (2021) Morphological variability of sandy beaches due to variable oceanographic conditions: a study case of oceanic beaches of Rio de Janeiro city (Brazil). J Coast Conserv 25:28

Coco G, Ciavola P (2017) Coastal storms. Wiley, Chichester, UK

Criado-Sudau FF, Nemes DD, Gallo MN (2019) Rip currents dynamic of a swell dominated microtidal beach. J Coast Res 121–127. https://doi.org/10.2112/SI92-014.1

Dadalto TP, Carvalho BC, Guerra JV et al (2021) Holocene morpho-sedimentary evolution of Marambaia Barrier Island (SE Brazil). Quat Res. https://doi.org/10.1017/qua.2021.43

da Silva Amador E (1987) Geologia e geomorfologia da planíciecosteira da Praia do Sul: Ilha Grande - uma contribuição à elaboração do plano diretor da reserva biológica. Anuário do Inst Geociências - UFRJ 11:35–58

da Silva ALC, Silva MAM, Santos CL (2008) Comportamento morfológico e sedimentar da praia de Itaipuaçú (Maricá, RJ) nas últimas três décadas. Rev Bras Geociências 38:87–97

da Silva AL, da Silva MA, Gralato JD, Silvestre CP (2014) Caracterização geomorfológica e sedimentar da planície costeira de Maricá (Rio de Janeiro). Rev Bras Geomorfol 15. https://doi.org/10.20502/rbg.v15i2.470

Dereczynski CP, Menezes WF (2015) Meteorologia da Bacia de Campos. In: Martins RP, Grossmann-Matheson GS (eds) Meteorologia e Oceanografia. Elsevier Editora Ltda, Rio de Janeiro, pp 1–54

Dietz M, Liu K, Bianchette T (2018) Hurricanes as a major driver of coastal erosion in the Mississippi River Delta: a multi-decadal analysis of Shoreline Retreat Rates at Bay Champagne, Louisiana (USA). Water 10:1480. https://doi.org/10.3390/w10101480

Dolan R, Davis RE (1992) An intensity scale for Atlantic Coast Northeast storms. J Coast Res 8:840–853

Eccard LR, da Silva ALC, Silvestre CP (2017) Morphological variations of ocean beaches of Niterói (RJ) in response to storms wave incidence. Rev Bras Geogr Física 10. https://doi.org/10.5935/1984-2295.20170011

Elko N, Dietrich C, Cialone M et al (2019) Advancing the understanding of storm processes and impacts. Shore Beach 87:37–50

Esteves LS, Williams JJ, Dillenburg SR (2006) Seasonal and interannual influences on the patterns of shoreline changes in Rio Grande do Sul, southern Brazil. J Coast Res 22:1076–1093. https://doi.org/10.2112/04-0227R.1

Fernandez GB, Bulhões E, da Rocha TB (2011) Impacts of severe storm occurred in April 2010 along Rio de Janeiro coast, Brazil. J Coast Res SI 64:1850–1854

Garnier E, Ciavola P, Spencer T et al (2018) Historical analysis of storm events: case studies in France, England, Portugal and Italy. Coast Eng 134:10–23. https://doi.org/10.1016/j.coastaleng.2017.06.014

Gentile DC, Goya SC (2005) Análise Morfodinâmica Da Praia Do Tombo, Guarujá-Sp. In: XIII Congresso da Associação Brasileira de Estudos do Quaternário. ABEQUA, Armação de Búzios, pp 1–5

Gentil de Farias EG, Parente Maia L (2010) Uso de Técnicas de Geoprocessamento para a Análise da Evolução da Linha de Costa em Ambientes Litorâneos do Estado do Ceará, Brasil. Rev Gestão Costeira Integr J Integr Coast Zo Manag 10:521–544

Gordon BL, Barnard P, Gibbons H (2017) Severe erosion on U.S. West coast during 2015–16 El Niño. Sound Waves FY 2017:1–2

Gralato J, da CA (2016) Dinâmica costeira e vulnerabilidade do litoral oriental da Ilha Grande (Angra dos Reis - RJ) às ondas de tempestades. Universidade do Estado do Rio de Janeiro

Gregório M das N, Araújo TCM de, Mendonça FJB et al (2017) Mudanças posicionais da linha de costa nas praias do Pina e de Boa Viagem, Recife, PE, Brasil. Trop Oceanogr 45:44–61. https://doi.org/10.5914/tropocean.v45i1.15200

Guerra JV, Ogston AS, Sternberg RW (2006) Winter variability of physical processes and sediment-transport events on the Eel River shelf, northern California. Cont Shelf Res 26:2050–2072. https://doi.org/10.1016/j.csr.2006.07.002

Harari J, de Camargo R (1994) Simulação da propagação das nove principais componentes de maré na plataforma sudeste brasileira através de modelo numérico hidrodinâmico. Brazilian J Oceanogr 42:35–54. https://doi.org/10.1590/S1679-87591994000100003

Harley MD (2017) Coastal storm definition. In: Ciavola P, Coco G (eds) Coastal storms. Wiley, Chichester, UK, pp 1–21

Himmelstoss EA, Henderson RE, Kratzmann MG, Farris AS (2021) Digital shoreline analysis system (DSAS) version 5.1 user guide. U.S. Geological Survey Open-File Report 2021-1091. https://doi.org/10.3133/ofr20211091

Honeycutt MG, Crowell M, Douglas BC (2001) Shoreline-position forecasting: impact of storms, rate-calculation methodologies, and temporal scales. J Coast Res 17:721–730. https://doi.org/10.2307/4300223

Innocentini V, Oliveira FA, Prado SCSC (2003) Modelo de ondas aplicado ao caso 5–8 de maio de 2001. Rev Bras Meteorol 18:97–104

Komar PD, Miller MC (1975) On the comparison between the threshold of sediment motion under waves and unidirectional currents with a discussion of the practical evaluation of the threshold. J Sediment Petrol 45:362–367. https://doi.org/10.1306/212F6D66-2B24-11D7-8648000102C1865D

Komen GJ, Hasselmann S, Hasselmann K (1984) On the existence of a fully developed wind-sea spectrum. J Phys Oceanogr 14:1271–1285

Lima GD, Menezes WF, Dereczynski CP, Ceccopieri W (2006) Ressaca no Rio de Janeiro provocada por um ciclone intenso: um estudo de caso. In: XIV Congresso Brasileiro de Meteorologia. SBMET, Florianópolis, pp 1–7

Losada IJ, Reguero BG, Méndez FJ et al (2013) Long-term changes in sea-level components in Latin America and the Caribbean. Glob Planet Change 104:34–50. https://doi.org/10.1016/j.gloplacha.2013.02.006

Machado AA, Calliari LJ, Melo E, Klein AHF (2010) Historical assessment of extreme coastal sea state conditions in southern Brazil and their relation to erosion episodes. J Aquat Sci 5:277–286

Maluf V, Fernandez G, Martins C et al (2016) Características morfodinâmicas tridimensionais da praia da Massambaba, litoral centro-sul do estado do Rio de Janeiro. In: XI SINAGEO. UEM, Maringá, pp 1–9

McFeeters S (1996) The use of the normalized difference water index (NDWI) in the delineation of open water features. Int J Remote Sens 17:1425–1432

Melet A, Meyssignac B, Almar R, Le Cozannet G (2018) Under-estimated wave contribution to coastal sea-level rise. Nat Clim Change 8:234–239. https://doi.org/10.1038/s41558-018-0088-y

Mendoza ET, Jimenez JA, Mateo J (2011) A coastal storm intensity scale for the Catalan sea (NW Mediterranean). Nat Hazards Earth Syst Sci 11:2453–2462. https://doi.org/10.5194/nhess-11-2453-2011

Mentaschi L, Vousdoukas MI, Pekel JF et al (2018) Global long-term observations of coastal erosion and accretion. Sci Rep 8:1–11. https://doi.org/10.1038/s41598-018-30904-w

Muehe D, Corrêa CHT (1989) Dinâmica de praia e transporte de sedimentos na restinga da Maçambaba, RJ. Rev Bras Geociências 19:387–392

Muehe D, Valentini E (1998) O Litoral do Estado do Rio de Janeiro: uma caracterização físico-ambiental. FEMAR, Rio de Janeiro

Muehe D (2011) Erosão costeira - tendência ou eventos extremos? O litoral entre Rio de Janeiro e Cabo Frio, Brasil. Rev Gestão Costeira Integr 11:315–325. https://doi.org/10.5894/rgci282

Muehe D, Klumb-Oliveira LA (2014) Deslocamento da linha de costa versus mobilidade praial. Quat Environ Geosci 5:121–124. https://doi.org/10.5380/abequa.v5i2.35884

Muehe D, Lins-de-Barros F, Oliveira JF de, Klumb-Oliveira LA (2015) Pulsos Erosivos E Resposta Morfodinâmica Associada a Eventos Extremos Na Costa Leste Do Estado Do Rio De Janeiro. Rev Bras Geomorfol 16. https://doi.org/10.20502/rbg.v16i3.728

Nemes DD, Criado-Sudau FF, Gallo MN (2019) Beach morphodynamic response to a submerged reef. Water 11:1–20. https://doi.org/10.3390/w11020340

Palmeira ACP, de A, Camargo R, Palmeira RM de J (2012) Effect of wind discretization in the modeling sea wave of western South Atlantic. Rev Bras Meteorol 27:117–125

Parente CE, Nogueira ICM, Ribeiro EO (2015) Climatologia de ondas. In: Martins RP, Grossmann-Matheson GS (eds) Meteorologia e Oceanografia. Elsevier Editora Ltda, Rio de Janeiro, pp 55–96

Parise C, Calliari LJ, Krusche N (2009) Extreme storm surges in the South of Brazil: atmospheric conditions and shore erosion. Braz J Oceanogr 57:1–23. https://doi.org/10.1590/S1679-875920 09000300002

Paula DP, Farrapeira Neto C de A (2017) Resposta de uma praia arenosa a um evento de ressaca do mar: o caso da Praia do Icaraí (Caucaia, Ceará, Brasil). Ateliê Geográfico 11:184. https://doi.org/10.5216/ag.v11i2.39302

Pena J, do N, Lins-de-Barros FM (2015) Gestão de praias urbanas e efeitos de eventos de ressaca: estudo de caso das praias urbanas oceânicas da cidade do Rio de Janeiro. In: XI Encontro Nacional da ANPEGE. ANPEGE, Presidente Prudente, pp 8122–8133

Pereira T, Saliba A, Lessa AC et al (2018) Morfodinâmica praial e os episódios erosivos na praia da Macumba, cidade do Rio de Janeiro, RJ. In: Anais do XII Simpósio Nacional de Geomorfologia. Crato

Phantuwongraj S, Choowong M, Nanayama F et al (2013) Coastal geomorphic conditions and styles of storm surge washover deposits from Southern Thailand. Geomorphology 192:43–58. https://doi.org/10.1016/j.geomorph.2013.03.016

Pinto VCS, Silva ALC da, Silvestre CP, Antonio RVM (2015) Ambiente praial e a influência do arenito de praia na distribuição dos sedimentos em Jaconé (Maricá, RJ). In: XVI Simpósio Brasileiro de Geografia Física Aplicada. UFPI, Tereseina, pp 1618–1625

Plant N, Doran K, Stockdon H (2017) Examples of storm impacts on barrier Islands. In: Coastal storms. Wiley, Chichester, UK, pp 65–79

Puig M, Del Río L, Plomaritis TA, Benavente J (2016) Contribution of storms to shoreline changes in mesotidal dissipative beaches: case study in the Gulf of Cádiz (SW Spain). Nat Hazards Earth Syst Sci 16:2543–2557. https://doi.org/10.5194/nhess-16-2543-2016

Rodrigues EM, Carvalho da Silva AL, Rodrigues Eccard L et al (2015) Vulnerabilidade e variações de curto prazo da praia de Itaipu (Niterói-RJ) em resposta às mudanças nas condições de mar. Rev Tamoios 10. https://doi.org/10.12957/tamoios.2014.13511

Roelvink D, Reniers A, van Dongeren A et al (2009) Modelling storm impacts on beaches, dunes and barrier islands. Coast Eng 56:1133–1152. https://doi.org/10.1016/j.coastaleng.2009.08.006

Santos CL, Silva MAM, Salvador MVS et al (2004) Dinâmica sazonal e os efeitos das ressacas nas praias de Niterói/RJ. Rev Bras Geociências 34:355–360
Splinter KD, Turner IL, Reinhardt M, Ruessink G (2017) Rapid adjustment of shoreline behavior to changing seasonality of storms: observations and modelling at an open-coast beach. Earth Surf Process Landforms 42:1186–1194. https://doi.org/10.1002/esp.4088
Storlazzi CD, Reid JA (2010) The influence of El Niño-Southern Oscillation (ENSO) cycles on wave-driven sea-floor sediment mobility along the central California continental margin. Cont Shelf Res 30:1582–1599. https://doi.org/10.1016/j
Wiles E, Loureiro C, Cawthra H (2022) Shoreline variability and coastal vulnerability: Mossel Bay, South Africa. Estuar Coast Shelf Sci 268:107789. https://doi.org/10.1016/j.ecss.2022.107789
Yokoyama CK (2012) Mobilidade sedimentar da plataforma continental do estado de São Paulo em função da propagação de ondas. Universidade de São Paulo
Young IR, Ribal A (2019) Multiplatform evaluation of global trends in wind speed and wave height. Science (80-) 9527:eaav9527. https://doi.org/10.1126/science.aav9527

Chapter 6
Geomorphological Effects of Tropical Cyclones in Costa Rica, Central America

Adolfo Quesada-Román

Abstract Tropical cyclones have provoked intense devastation over the last decades, and societal drivers such as unplanned urbanization and the lack of flood risk assessment have worsened the problem. Landslides and floods constitute a frequent hazard linked to tropical cyclones. This is not the exception in Costa Rica, where hydrometeorological disasters sum more than 90% of the total. Different climatic change scenarios indicate that a large number and more strong tropical cyclones will affect the North Atlantic Ocean Basin, where Central America and the Caribbean Sea lie. This study comprises different methodologies to understand the geomorphic effects of tropical cyclones in Costa Rica. This chapter presents geomorphological, hydrological, dendrochronological, and risk assessment approaches that aim to reduce the human and economic impact of tropical cyclones. Finally, the chapter calls for applying geomorphic assessments to elucidate the endogenic and exogenic processes controlling the effects of tropical cyclones in low latitudes.

Keywords Tropical geomorphology · Disaster risk reduction · Developing countries · Caribbean Sea · Latin America · Global South

6.1 Introduction

Climate change will raise the frequency and occurrence of high-intensity tropical cyclones. This phenomenon is projected to fold up the impacts and losses, particularly in East Asia, North America, and the Caribbean–Central American regions (Mendelsohn et al. 2012; Pinos and Quesada-Román 2022). Several climatic modes control low-latitude dynamics, i.e., the ITCZ, the trade winds, cold fronts, ENSO, cyclonic systems, and the orographic effect (Wohl et al. 2012). Landslides, floods, tropical storms, and droughts constitute the most common hydrometeorological hazards in

---

A. Quesada-Román (✉)
Universidad de Costa Rica, Escuela de Geografía, Laboratorio de Geografía Física, 2060 San José, Costa Rica
e-mail: adolfo.quesadaroman@ucr.ac.cr

A. Coronato and G. B. Alves (eds.), *Latin American Geomorphology*, The Latin American Studies Book Series, https://doi.org/10.1007/978-3-031-55178-9_6

the tropics (UNDRR 2019). Floods have caused a significant amount of the all-natural catastrophic economic losses over the last decades globally (Aerts et al. 2018; IPCC 2014). In addition, most natural disasters in tropical countries originate due to hydrometeorological events, including landslides and floods (Alcántara-Ayala 2002).

The occurrence of extreme climatic events and flood frequency has increased lately, whereas natural forest cover has declined across the tropics (Lawrence and Vandecar 2015). Tropical floods are controlled by the interconnection of land use changes, floodplain occupation, and extreme variable climatological events (Syvitski et al. 2014). Accelerated deforestation rates in tropical rainforests negatively impact local public health, agricultural productivity, and global climate change (Vargas-Zeppetello et al. 2020). The Atlantic Multidecadal Oscillation (AMO) relates with the sea surface temperature (SST) variations in the tropics (Knight et al. 2006; Kayano and Capistrano 2014; Sun et al. 2017). Hence, ENSO conditions are enhanced, and monsoon or tropical cyclones appear more frequently (Wang et al. 2013). Extreme La Niña and El Niño events control the occurrence of intense dry or wet conditions along the tropics (NOAA 2022). The effect of ENSO denotes that throughout the La Niña phase, there are more possibilities of having tropical storms as opposed to experiencing an El Niño phase (Goldenberg et al. 2001). Therefore, tropical greater contributions and quicker rates of change have recently caused more extreme floods (Wohl et al. 2012).

There is a clear statistical trend since the 1970s in the major hurricanes number in the Caribbean Sea (Alfaro and Quesada-Román 2010; Alfaro et al. 2010). Out of all the cyclones that developed in the Atlantic basin, 14% generated indirect impacts in Costa Rica, but the probability for a direct impact was below 6% in the twentieth century (Alvarado and Alfaro 2003). Costa Rica is normally impacted indirectly by hurricanes, particularly in the Pacific basin, but until 2016 when Hurricane Otto hit the country, none of these cyclones directly impacted (Quesada-Román and Villalobos-Chacón 2020). During the rainy season, smaller disasters occur regularly whereas tropical cyclones are typically the cause of extraordinary floods and landslides in Costa Rica (Quesada-Román et al. 2021a, b).

Every year, Costa Rica uses roughly 1% of its Gross Domestic Product (GDP) to repair the damage that results because of disasters (Astorga 2010). Disaster impacts illustrate the escalating vulnerability of the country. The recovery cost of recent tropical cyclones has been significant. Hurricane Otto (2016) represented 0.4% and tropical storm Nate (2017), up to 1.3% of the GDP. Moreover, hurricane Eta (2020) was worth more than 1.6 million USD in economic impacts. These disastrous consequences were predominantly related to fixing bridges and roads (road infrastructure) where an evident lack exists (Brenes and Girot 2018).

The aim of this chapter is to depict recent endeavors to reduce the impact of tropical cyclones in Costa Rica, using different Earth and Social sciences methods. We hypothesize that different geomorphological, hydrological, dendrochronological, and risk assessment approaches could prove successful to reduce the socioeconomic impacts of tropical cyclones in Costa Rica in the next decades. Moreover, the chapter might be an example for developing/tropical regions or countries facing extraordinary rainfall events as a consequence of climate change and higher exposure in their territories.

6.2 Costa Rica's Geographical Setting

Tectonic activity of Costa Rica is composed by the interaction of the Cocos and Caribbean plate's subduction margin, the microplate of Panama, and the subduction of the Cocos volcanic range (Alvarado et al. 2017). The country is divided into three morphotectonic units: a forearc, a volcanic front, and a backarc that comprise the country's lithology (Marshall 2007). The forearc widens along the Pacific seaside with intricate morphologies since the Cretaceous age (Denyer and Alvarado 2007). The volcanic front comprises the main cordilleras made mostly of volcanic and sedimentary rocks since the Paleogene age (Alvarado 2021). The backarc spreads from the Caribbean plains of Tortuguero's flatlands to the hilly relief of the Caribbean south (Quesada-Román and Pérez-Briceño 2019). The principal geomorphic dynamics in the country are fluvial, volcanic, and coastal, but some karstic and glacial areas also pop up (Quesada-Román and Pérez-Umaña 2020).

The continuous sequence of cordilleras that cross over Costa Rica with an NW–SE alignment characterizes the Pacific and Caribbean basins (Fig. 6.1). This topographic barrier also controls the amount of rainfall in each basin: the Pacific and Caribbean basins (Amador et al. 2010; Quesada-Román et al. 2020b). The Pacific Ocean climate shows a bimodal precipitation regime, but it is hard to identify a dry period in the Caribbean Sea climate (Maldonado et al. 2018; Castillo and Amador 2020). The Caribbean side's annual rainfall totals are up to 3000 mm; they are generally below 3 m in the Pacific (Fig. 6.2). These climatological characteristics justify that over 90% of disasters in the country are of hydrometeorological origin. Moreover, 60% of them are floods and 30% are mass movements (Campos-Durán and Quesada-Román 2017; LA RED 2022).

The high number of endemic species, levels of forest composition and configuration, and a varied biodiversity position Costa Rica as a megadiverse country. The country holds approximately 6% of the worldwide biodiversity, consisting of mountain cloud forests, evergreen moist forests, seasonal forests, dry forests, paramos, coastal, and wetlands ecosystems. The country's population reached 5 million people in 2018, and, over the last decades, its dynamics changed from a significant rurality to a strong urban trend achieving three quarters of the inhabitants in 2011. At present, the Greater Metropolitan Area (or GAM) has roughly 3 million inhabitants in only 14% of the country's land (Van Lidth de Jeude et al. 2016). This urban sprawl poses the GAM as an exposed and vulnerable region to the impacts of landslides and floods every year (Quesada-Román et al. 2021b). Mass movement and flooding research have been vast in different regions of the country but published mostly in Spanish (Quesada-Román 2021, 2022). Nevertheless, the production of detailed-scale hazard maps and its combination with vulnerability and risk calculation is scarce (Quesada-Román 2017; Quesada-Román and Zamorano-Orozco 2019b).

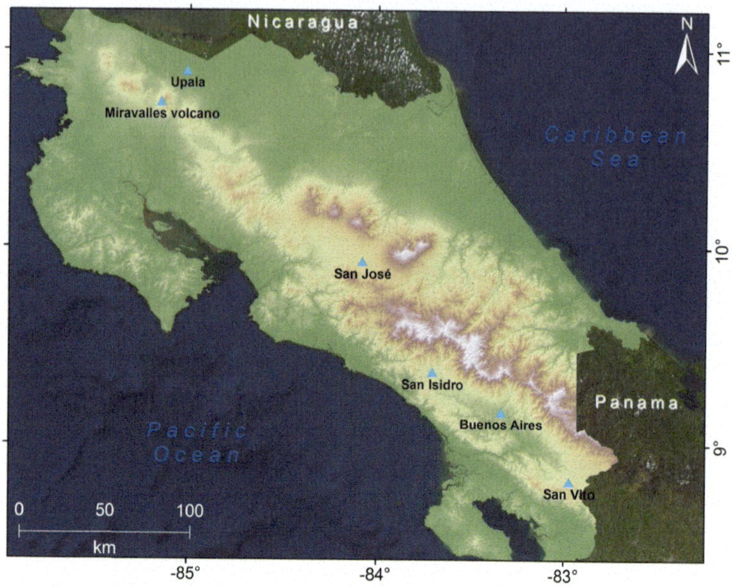

Fig. 6.1 Location of the main study sites of the chapter

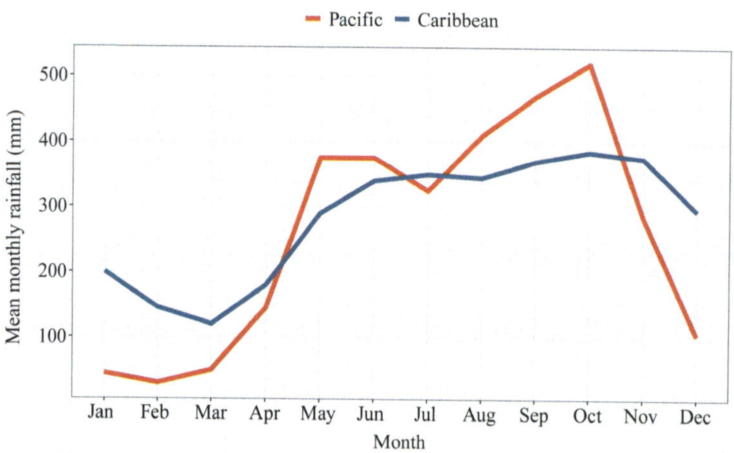

Fig. 6.2 Rainfall trends in the Pacific and Caribbean basins of Costa Rica

6.3 Methods and Techniques to Analyze the Impacts of Tropical Cyclones

6.3.1 Geomorphological Hazards Mapping

The General River catchment is a very dynamic water basin located 85 km SE of the city of San José where San Isidro del General is the main city of the region with approximately 40,000 inhabitants. Here, the catchment is defined by the interaction of the Cordillera de Talamanca, the General River valley, and the Fila Brunqueña units in a territory of 1,560 km^2. Different cyclones historically have impacted the region.

This work was based on pre-mapping, fieldwork, and post-mapping phases (Smith et al. 2011). Throughout the pre-mapping, geomorphometry and morphogenetic mapping were produced using aerial photos (1:25,000). Much fieldwork was made between 2011 and 2015 to verify the landform dynamics and their limits using initial maps. The morphometric assessment evaluated seven parameters (dissection density, altimetry, relief energy, total erosion, slopes, dissection profundity, and potential erosion; Quesada-Román and Barrantes 2017; Quesada-Román and Feoli-Boraschi 2018).

The slopes map is the first approximation of the gravitational dynamics. It identifies landforms linked with the fluvial erosive/accumulative dynamics. The dissection density shows zones of greater/lesser erosion, where the channel's concentration is linked to rainy zones, unconsolidated and weathered sediments, disjunctive structures, large areas without significant lithological changes, and heterogeneous morphology. Dissection profundity makes clear the influence of the disjunctive structures in the erosive and accumulative relief dynamics. The relief energy combines fluvial (dissection density and dissection profundity) and slope action in its analysis. Potential erosion locates indirectly areas with denudation susceptibility, linked to factors such as specific lithological characteristics, topographic contrasts, morphological variety, hydroclimatic conditions, vegetation cover, and land use changes. Finally, total erosion determines areas with greater erodibility and with older morphologies were distinguished through the terrain roughness, and the role played by variables such as lithology age, disjunctive structures, the past climate (glaciations), the current rainfall patterns, and slopes.

These parameters were statistically combined to generate the flooding and mass movement susceptibility cartography, categorized into three classes (high, medium, and low) using the Natural Jenks classification method. This procedure permitted a parallel evaluation using the disaster database DesInventar (LA RED 2022). Post-mapping phase comprises the geomorphological map legend elaboration, separating the landforms into endogenic and exogenic processes, and erosional and depositional dynamics. All mappings were made within a Geographic Information System (ArcGIS 10.8).

Using geomorphometry as an indicator, the territories can be considered potential, frequent, and maximum occurrence areas for landslides and floods using a frequency

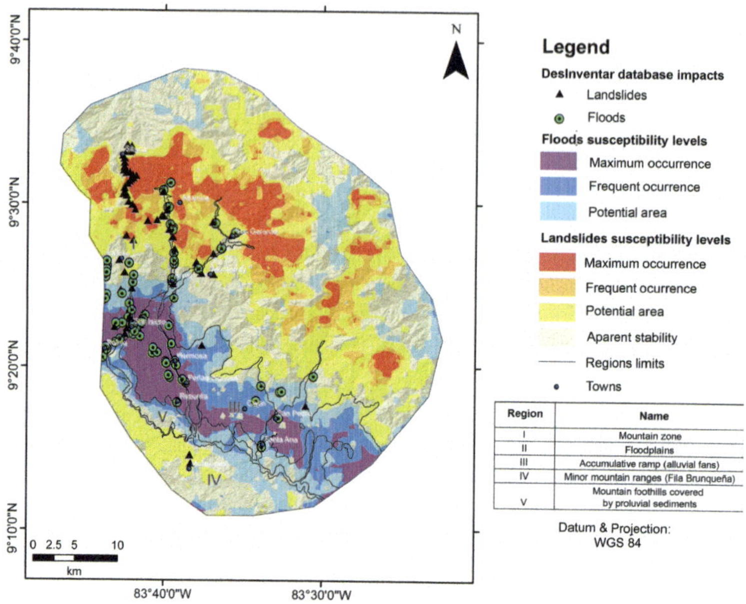

Fig. 6.3 Susceptibility zonation of mass movements and flooding using geomorphometry

analysis between the used variables (Quesada-Román and Zamorano-Orozco 2019b). The maps were geographically compared with the DesInventar database (Quesada-Román and Zamorano-Orozco 2018; LA RED 2022). An 85% of the reported events agree with the identified prone flooding zones, and 76% with mass wasting vulnerable areas (Fig. 6.3). Moreover, Quesada-Román (2016) and Quesada-Román and Zamorano (2019a), mapped 46 morphologies, arranged in endogenic (tectonic) and exogenic (fluvial, gravitational, and glacial) morphologies. A geomorphological hazard map was created merging geomorphometry and morphogenetic maps, which helped to determine distinct susceptibility classes for floods and landslides. We merged active landslides and flood areas with high geomorphometry values to create this map. Landslides displayed maximum occurrence, low occurrence, and potential area susceptibility classes. Floods were classified as apparent bed (seasonal), maximum floodplain, extraordinary floodplain, and low susceptibility to occurrence (Figs. 6.4 and 6.5).

6.3.2 *Can Earthquakes and Hurricanes Merge?*

The Miravalles volcano, in northern Costa Rica, was impacted by an earthquake of 5.4 Mw (July 2016) and followed by extraordinary precipitations associated with Hurricane Otto (category 3) just a few weeks after (November 2016). Hurricane

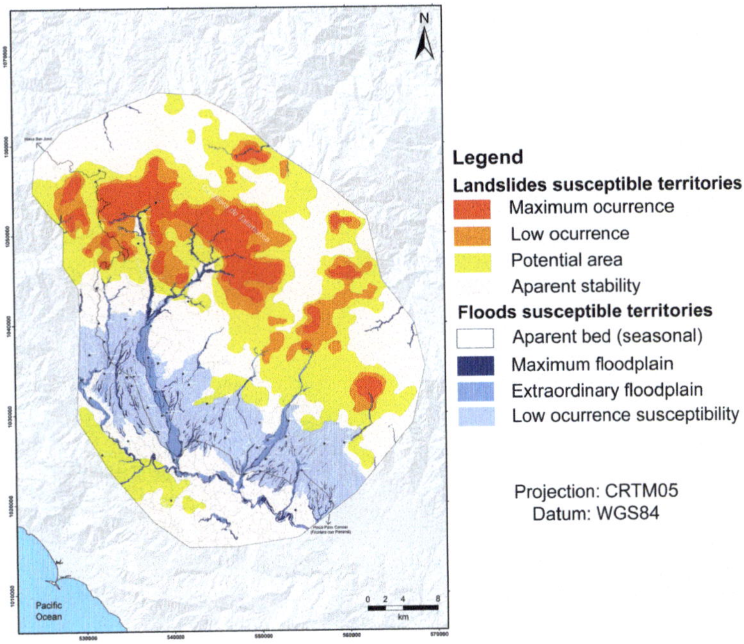

Fig. 6.4 Mass movements and flooding zonation

Otto's passing left roughly 300 mm of rain over the volcano. Using a general linear regression model (GLM), it was possible to characterize the interactions among the parameters influencing landslides (epicenter distance, precipitation, altimetry, and slope).

The cascading 2016 disaster caused 942 mass movements (Fig. 6.6). 62% of them occurred between 3 and 6 km from the prior epicenter and in the E, SE, and S slopes of Miravalles volcano where local faults density concentrate, and precipitation summed maximum values throughout the cyclone (Quesada-Román et al. 2019). The GLM validates a combined earthquake-hurricane coupling with greater mass movement densities near the epicenter, at zones with greater rainfall amount, in steep and higher regions.

The application of an S1 Ground Range Detected and VV polarization to Sentinel-1 synthetic aperture radar and WorldView-3 and -4 images helped to map the flash-flooded areas after Hurricane Otto's impact in Upala municipality (Fig. 6.7). Debris flows impacted roughly 27 km^2 (Fig. 6.8) and moved down the slopes as flash floods in approximately 74 km^2 (Fig. 6.9a, b), with eight human casualties in the vicinity of the Miravalles volcano and roughly 103 million US$ of losses in the Bagaces and Upala municipalities.

Fig. 6.5 **a** Mass movements maximum occurrence over Pan-American highway. **b** Built dams in a seasonal floodplain in San Isidro del general

6.3.3 Dendrogeomorphology and Disaster Risk Assessments

The Térraba catchment in central-south has the bigger area and peak discharge records in Costa Rica with 4,765 km^2 (Fig. 6.12). Seven random sample places and hydrodynamic models to calculate peak discharges along the catchment were used to reconstruct the peak discharge of a recent tropical cyclone (Tropical Storm Nate 2017) employing botanical evidence, especially scars on the trees (Fig. 6.10). Interestingly, trees on cut banks or terraces have minimal uncertainties among observed and modeled heights of flows, not in straight channel reaches (Quesada-Román et al. 2020a; Fig. 6.11).

In the second phase, the available data from hydrological data of yearly maximum flows from eight stations from 1962 to 2019 was employed to apply a regional flood-frequency method employing the Bayesian Markov Monte Carlo Chain (MCMC) algorithm (Gaál et al. 2010; Gaume 2018). Consequently, regional flood-frequency analyses were performed. The study combined a dendrochronology assessment and peak discharge calculation to obtain flood quartiles. These quartiles were merged with the Topographic Wetness Index (TWI) to ascertain regional flood hazards along

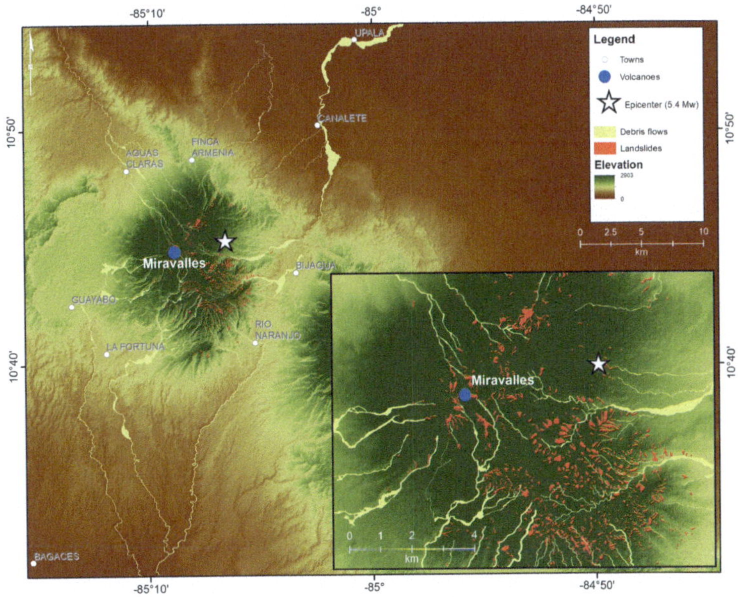

Fig. 6.6 Debris flows and landslides generated by Hurricane Otto over Miravalles volcano

Fig. 6.7 **a** Aerial views over Miravalles volcano following the cyclone. **b** Mass movement on the Miravalles volcano NE slope, the photo is heading to the SW. **c** Large blocks and wood moved along the NE volcano slope. **d** The debris flows produced four human casualties in Bijagua de Upala and immense economic losses of over 100 million US$

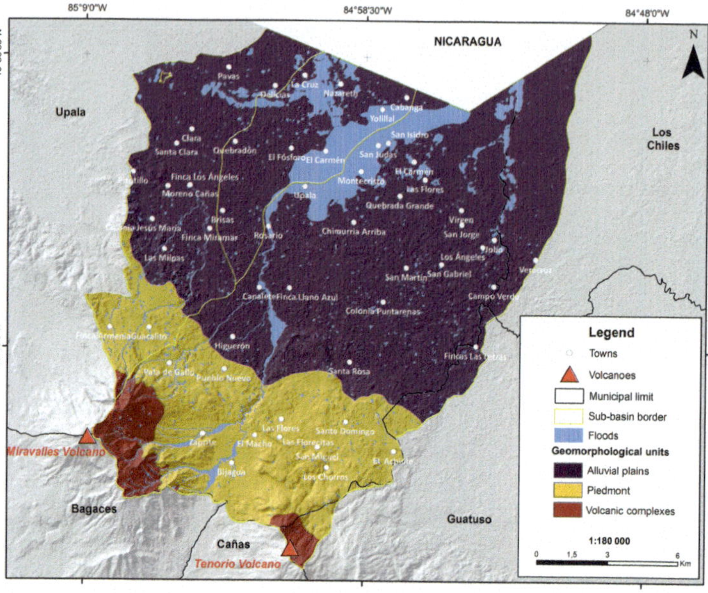

Fig. 6.8 Geomorphological units and Hurricane Otto flash floods in Upala municipality

Fig. 6.9 Aerial views over Upala after the passing of Hurricane Otto. **a** Debris flows and flash floods carried large wood. **b** Downstream the floodplains spread out. **c**, **d** Rice and pineapple croplands and forest/wetlands fragmented areas were the most flooded areas

Fig. 6.10 Successive inlet modeled discharges using historical extremes, dendrogeomorphology, and a two-dimensional hydrodynamic model to calculate peak discharges. Figure is not escalated

riverbeds and surrounding areas (Fig. 6.12). Flood exposure was determined using the population and infrastructure density (Tatem 2017; Fig. 6.12). Flood vulnerability was based on the social development index (IDS 2017) which merges 14 socioeconomic parameters nationally (Fig. 6.12; MIDEPLAN 2017). The flood risk assessment (Fig. 6.12) merges the regional flood-frequency analysis coupled with dendrochronology (hazard), a detailed cartography of population/infrastructure density (exposure), and a social development index (vulnerability). The transdisciplinary approach combination proved useful in the scarce-data Térraba catchment and about 6,000 people live in flooding risky areas (Quesada-Román et al. 2022).

6.4 Conclusions

Several geomorphological techniques were presented to study the impacts of tropical cyclones in Costa Rica. Geomorphometry is a handy method for disaster risk mitigation determining probable different degrees of impact of mass movements and flooding. Nevertheless, geomorphometry needs a detailed adjustment using the morphogenetic mapping to limit the different susceptible areas in detail. Geomorphometry assessments can be the reference line for integrated catchment management and municipalities at regional and local levels. Previously affected mountain regions by earthquakes must be monitored in extraordinary precipitation events to

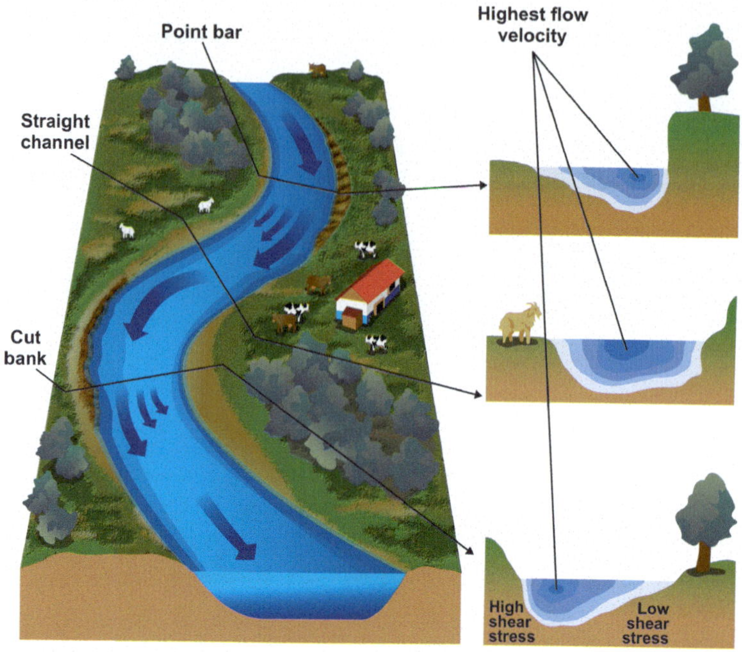

Fig. 6.11 Tree's geomorphic location controls the uncertainties among observed (scars) and calculated peak discharge. Uncertainties are lower on stable fluvial landforms such as point bars, terraces, and cut banks

anticipate further disasters. Areas, where mass movements generated by earthquakes and subsequent strong rainfall events can be triggered, are necessary to be studied to minimize casualties and losses in developing countries. Flood-frequency analyses with dendrogeomorphic estimations and hydrological measurements can be precisely used to calculate cyclone-induced peak discharges and floods. Coarse and detailed data can be useful implementing regional flood risk evaluations in large-scale catchments. A regional flood-frequency approach in parallel with dendrogeomorphology and risk assessment is a practical contribution for territorial planning strengthening the resilience of tropical catchment populations. This multidisciplinary procedure can be employed in countries where hydrological reference information is limited.

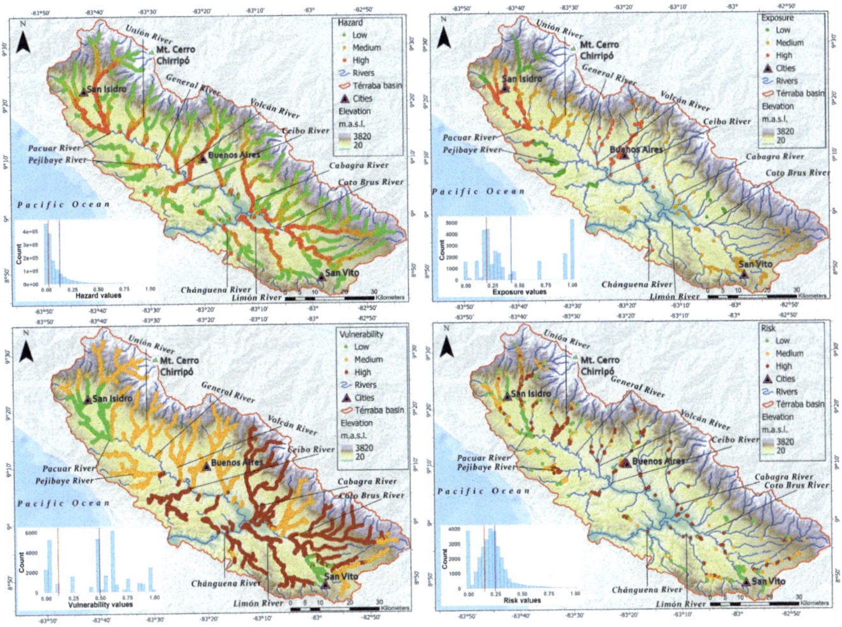

Fig. 6.12 Flood hazard (upper left), exposure (upper right), vulnerability (lower left), and risk (lower right) maps of the Térraba catchment, Costa Rica

Acknowledgements To Soll Kracher, anonymous reviewers, and editors who highly improved the final version of the manuscript.

References

Aerts JC, Botzen WJ, Clarke KC, Cutter SL, Hall JW, Merz B, Michel-Kerjan E, Mysiak J, Surminski S, Kunreuther H (2018) Integrating human behaviour dynamics into flood disaster risk assessment. Nat Clim Change 8(3):193–199

Alcántara-Ayala I (2002) Geomorphology, natural hazards, vulnerability and prevention of natural disasters in developing countries. Geomorphology 47(2–4):107–124

Alfaro E, Quesada-Román A (2010) Ocurrencia de ciclones tropicales en el Mar Caribe y sus impactos sobre Centroamérica. Revista InterSedes 11(22):136–153

Alfaro E, Quesada-Román A, Solano FJ (2010) Análisis del impacto en Costa Rica de los ciclones tropicales ocurridos en el Mar Caribe desde 1968 al 2007. Revista Diálogos 11(2):25–38

Alvarado LF, Alfaro E (2003) Frecuencia de los ciclones tropicales que afectaron a Costa Rica durante el siglo XX. Tóp Meteo Ocean 10(1):1–11

Alvarado GE, Benito B, Staller A, Climent A, Camacho E, Rojas W, Marroquín G, Molina E, Talavera JE, Martínez-Cuevas S, Lindholm C (2017) The new Central American seismic hazard zonation: mutual consensus based on up to day seismotectonic framework. Tectonophysics 721:462–476

Alvarado GE (2021) Costa Rica y sus volcanes. San José, Costa Rica: EUCR, EUNA, ETCR

Amador JA, Alfaro EJ, Rivera ER, Calderón B (2010) Climatic features and their relationship with tropical cyclones over the Intra-Americas seas. In: Hurricanes and climate change. Springer, Dordrecht, pp 149–173
Astorga A (2010) Ordenamiento territorial en Costa Rica. Decimoséptimo Estado de la Nación, CONARE, Costa Rica
Brenes A, Girot P (2018) Gestión del riesgo y cambio climático. Informe Estado de la Nación en Desarrollo Humano Sostenible. CONARE, Programa Estado de la Nación, Pavas, Costa Rica, pp 1–52
Campos-Durán D, Quesada-Román A (2017) Impacto de los eventos hidrometeorológicos en Costa Rica, periodo 2000–2015. Revista Geo UERJ 30:440–465
Castillo R, Amador JA (2020) Precipitation and temperature in Costa Rica at the end of the century based on NEX-GDDP projected scenarios. Atmosphere 11(12):1323
Denyer P, Alvarado GE (2007) Mapa geológico de Costa Rica. Escala 1:400 000. Librería Francesa. San José, Costa Rica
Gaál L, Szolgay J, Kohnová S, Hlavčová K, Viglione A (2010) Inclusion of historical information in flood frequency analysis using a Bayesian MCMC technique: a case study for the power dam Orlík, Czech Republic. Contrib Geophys 40(2):121–147
Gaume E (2018) Flood frequency analysis: the Bayesian choice. Wires Water 5(4):e1290
Goldenberg SB, Landsea CW, Mestas-Nuñez AM, Gray WM (2001) The recent increase in Atlantic hurricane activity: causes and implications. Science 293(5529):474–479
IPCC (2014) In: Field et al (eds) Climate change 2014: impacts, adaptation, and vulnerability. Part A: global and sectoral aspects. Contribution of Working Group II to the Fifth Assessment Report of the Intergovernmental Panel on Climate Change (IPCC). Cambridge University Press, Cambridge, UK and New York, USA, p 1132
Kayano MT, Capistrano VB (2014) How the Atlantic multidecadal oscillation (AMO) modifies the ENSO influence on the South American rainfall. Int J Climatol 34(1):162–178
Knight JR, Folland CK, Scaife AA (2006) Climate impacts of the Atlantic multidecadal oscillation. Geophys Res Lett 33(17):1–4
LA RED (2022) DesInventar—Inventory system of the effects of disasters. Corporación OSSO, LA RED, UNDRR. http://desinventar.org. Accessed 15 Mar 2022
Lawrence D, Vandecar K (2015) Effects of tropical deforestation on climate and agriculture. Nat Clim Change 5(1):27–36
Maldonado T, Alfaro E, Hidalgo HG (2018) A review of the main drivers and variability of Central America's Climate and seasonal forecast systems. Revista Biol Trop 66(1–1):S153–S175
Marshall J (2007) The Geomorphology and physiographic provinces of central America. In Bundschuh J, Alvarado GE (eds) Central America: geology, resources and hazards. Taylor & Francis, p 1436
Mendelsohn R, Emanuel K, Chonabayashi S, Bakkensen L (2012) The impact of climate change on global tropical cyclone damage. Nat Clim Change 2(3):205–209
MIDEPLAN—Ministerio de Planificación Nacional y Política Económica (2017) Índice de Desarrollo Social 2017. San José, Costa Rica, p 126
NOAA—National Oceanic and Atmospheric Administration (2022) El Niño and La Niña years and intensities based on Oceanic Niño Index (ONI). https://origin.cpc.ncep.noaa.gov/products/analysis_monitoring/ensostuff/ONI_v5.php. Accessed 15 Mar 2022
Pinos J, Quesada-Román A (2022) Flood risk-related research trends in Latin America and the Caribbean. Water 14(1):10
Quesada-Román A (2016) Peligros geomorfológicos: inundaciones y procesos de ladera en la cuenca alta del río General (Pérez Zeledón), Costa Rica. Maestría en Geografía con énfasis en Geografía Ambiental. Posgrado en Geografía. Universidad Nacional Autónoma de México, 157 pp
Quesada-Román A (2017) Geomorfología Fluvial e Inundaciones en la Cuenca Alta del Río General, Costa Rica. Anuá Inst Geociê 40(2):278–288

Quesada-Román A, Barrantes G (2017) Modelo morfométrico para determinar áreas susceptibles a procesos de ladera. Inv Geogr 94:1–12

Quesada-Román A, Feoli-Boraschi S (2018) Comparación de la metodología Mora-Vahrson y el método morfométrico para determinar áreas susceptibles a deslizamientos en la microcuenca del río Macho, Costa Rica. Revista Geogr América Central 2(61):17–45

Quesada-Román A, Zamorano-Orozco JJ (2018) Peligros Geomorfológicos en Costa Rica: Cuenca Alta del Río General. Anuá Inst Geociê 41(3):239–251

Quesada-Román A, Moncada-López R, Paz-Tenorio JA, Espinoza-Jaime E, Castellón-Meyrat C, Acosta-Galeano N (2018) Las investigaciones sobre movimientos de laderas en Costa Rica, Honduras, México y Nicaragua: enseñanzas desde la academia, las agencias de cooperación y las instituciones públicas. Revista Geogr Amér Cent 60:17–59. https://doi.org/10.15359/rgac.60-1.1

Quesada-Román A, Pérez-Briceño PM (2019) Geomorphology of the Caribbean coast of Costa Rica. J Maps 15(2):363–371

Quesada-Román A, Zamorano-Orozco JJ (2019a) Geomorphology of the upper general River Basin. Costa Rica. J Maps 15(2):95–101

Quesada-Román A, Zamorano-Orozco JJ (2019b) Zonificación de procesos de ladera e inundaciones a partir de un análisis morfométrico en la cuenca alta del río General, Costa Rica. Inv Geog 99:1–19

Quesada-Román A, Fallas-López B, Hernández-Espinoza K, Stoffel M, Ballesteros-Cánovas JA (2019) Relationships between earthquakes, hurricanes, and landslides in Costa Rica. Landslides 16(8):1539–1550

Quesada-Román A, Pérez-Umaña D (2020) State of the art of geodiversity, geoconserva-tion, and geotourism in Costa Rica. Geosciences 10(6):211

Quesada-Román A, Villalobos-Chacón A (2020) Flash flood impacts of Hurricane Otto and hydrometeorological risk mapping in Costa Rica. Geogr Tidsskr 120(2):142–155

Quesada-Román A, Ballesteros-Cánovas JA, Granados-Bolaños S, Birkel C, Stoffel M (2020a) Dendrogeomorphic reconstruction of floods in a dynamic tropical river. Geomorphology 359:107133

Quesada-Román A, Ballesteros-Cánovas JA, Guillet S, Madrigal-González J, Stoffel M (2020b) Neotropical Hypericum irazuense shrubs reveal recent ENSO variability in Costa Rican páramo. Dendrochronologia 61:125704

Quesada-Román A (2021) Landslide risk index map at the municipal scale for Costa Rica. Int J Disaster Risk Reduct 56:102144

Quesada-Román A, Castro-Chacón JP, Feoli-Boraschi S (2021a) Geomorphology, land use, and environmental impacts in a densely populated urban catchment of Costa Rica. J S Am Earth Sci 112:103560

Quesada-Román A, Villalobos-Portilla E, Campos-Durán D (2021b) Hydrometeorological disasters in urban areas of Costa Rica, Central America. Environ Hazards 20(3):264–278

Quesada-Román A (2022) Flood risk index development at the municipal level in costa rica: a methodological framework. Environ Sci Policy 133:98–106

Quesada-Román A, Ballesteros-Cánovas JA, Granados-Bolaños S, Birkel C, Stoffel M (2022) Improving regional flood risk assessment using flood frequency and dendrogeomorphic analyses in mountain catchments impacted by tropical cyclones. Geomorphology 396:108000

Smith MJ, Paron P, Griffiths JS (2011) Geomorphological mapping: methods and applications, vol 15. Elsevier, p 612

Sun C, Kucharski F, Li J, Jin FF, Kang IS, Ding R (2017) Western tropical pacific multidecadal variability forced by the Atlantic multidecadal oscillation. Nature Commun 8(1):1–10

Syvitski JPM, Cohen S, Kettner AJ, Brakenridge GR (2014) How important and different are tropical rivers? An overview. Geomorphology 227:5–17

Tatem AJ (2017) WorldPop, open data for spatial demography. Sci Data 4(1):1–4

UNDRR—United Nations Disaster Risk Reduction (2019) Global assessment report on disaster risk reduction. United Nations Office for Disaster Risk Reduction (UNDRR). Geneva, Switzerland

Van Lidth de Jeude M, Schütte O, Quesada F (2016) The vicious circle of social segregation and spatial fragmentation in Costa Rica's greater metropolitan area. Habitat Int 54:65–73
Vargas-Zeppetello L, Parsons L, Spector J, Naylor R, Battisti D, Masuda Y, Wolff NH (2020) Large scale tropical deforestation drives extreme warming. Environ Res Lett 15(8):84012
Wang B, Liu J, Kim HJ, Webster PJ, Yim SY, Xiang B (2013) Northern Hemisphere summer monsoon intensified by mega-El Niño/southern oscillation and Atlantic multidecadal oscillation. Proc Natl Acad Sci USA 110(14):5347–5352
Wohl E, Barros A, Brunsell N, Chappell NA, Coe M, Giambelluca T, Goldsmith S, Harmon R, Hendrickx JMH, Juvik J, McDonnell J, Ogden F (2012) The hydrology of the humid tropics. Nat Clim Change 2(9):655–662

Chapter 7
Illumination Conditions at Mars and Their Relationships with Ice-Driven Morphology

Mauro G. Spagnuolo, Mara Mantegazza, and Santiago H. Luna

Abstract Geomorphology as a discipline has extended its horizons to reach other celestial bodies. Planets, dwarf planets, moons, and asteroids can be studied by applying the same geomorphological premises. Through indirect information, planetary surfaces can be interpreted in a comparative way, applying the principles of terrestrial geomorphology to other planets but taking into account the physical properties of each planet. Nevertheless, some caution must be taken since a similar landform can have different origins on different planets. Mars has several geomorphic features that might be related with seasonal changes and liquid water/ice interaction. In this chapter, we make use of the MarsLux code to investigate changes in illumination conditions and relate the illumination and insolation maps with different landforms. For example, we could relate the resulting maps with cycles of freezing and thawing on the Martian surface. Also, by generating maps of illumination over one Martian year, we can relate the location of certain features with specific areas that receive an anomalous amount of solar energy. We present a tool that can connect an illumination analysis, from orbital parameters, with changes in insolation over geologically recent times and geomorphological evidence that points to the presence of subsurface ice, or alternatively to the thermal weathering of rocks.

Keywords Mars · Illumination · Geomorphology · Volitiles

M. G. Spagnuolo (✉) · M. Mantegazza
IDEAN (UBA-CONICET), Universidad de Buenos Aires, Intendente Güiraldes 2160, Ciudad Universitaria - Pabellón II, C1428EGA - CABA Buenos Aires, Argentina
e-mail: mauro@gl.fcen.uba.ar

M. Mantegazza
e-mail: mmantegazza@gl.fcen.uba.ar

S. H. Luna
UNAHUR, Teniente Origone 151, B1688AXC, Hurlingham, Buenos Aires, Argentina

IDEAN (UBA-CONICET), Ciudad Universitaria - Pabellón II, Intendente Güiraldes 2160, C1428EGA - CABA Buenos Aires, Argentina

A. Coronato and G. B. Alves (eds.), *Latin American Geomorphology*, The Latin American Studies Book Series, https://doi.org/10.1007/978-3-031-55178-9_7

7.1 Introduction

7.1.1 General Mars Geomorphology

Mars could be considered an active planet in terms of its geomorphology. Recent review papers by Diniega et al. (2021) and Dundas et al. (2021) provide an excellent and in-depth review of Martian geomorphology. Along with the review papers on atmospheric processes by Smith (2008) and Martínez et al. (2017), and the climate reviews by Diniega and Smith (2020) and Piqueux et al. (2015b), these works cover and summarize most of the active processes acting today on Mars. In this introduction, we highlight some of their main points.

The surface of Mars is being modified by several processes as a consequence of its interaction with wind, CO_2 and H_2O cycles, gravity (i.e., mass wasting), and insolation. Although insolation might only be responsible directly for very superficial processes, it can be considered a major driver behind the thermal state of the planet's surface. Not only atmospheric processes are in large part driven by changes in insolation (Piqueux et al. 2015a, b; Spiga and Forget 2009) but also insolation can indirectly cause "Marsquake-seasons" by changes in the topographic loads generated by ice/snow (Knapmeyer et al. 2021).

Seasonal variations on Mars are more extreme than on Earth, with huge year-to-year differences in atmospheric conditions. The most obvious difference is the fact that Martian surface pressure varies seasonally by ~30%, as the wintertime deposition of an ephemeral layer of CO_2 frost condenses out a significant fraction of the atmosphere (Leighton and Murray 1966; Smith 2008).

In terms of landforms, many similar geomorphic features are found on both planets (Earth and Mars). The Martian surface is covered by erosional features such as gullies, which resemble (and get their name) from their terrestrial counterparts' fluvial erosion cavities, depressions of diverse shapes resulting from a variety of processes, like sublimation, and mass-wasting processes (Fig. 7.1). Nevertheless, in terms of the physical processes that originate these features, the Martian environment can be unique. Mars' surface temperatures range from ~130 K at the poles to ~230 K at the equator (Kieffer 2013; Piqueux et al. 2016). The atmospheric pressure varies from >1,000 Pa at the bottom of the Hellas basin to <100 Pa at the summit of Olympus Mons. Condensation of CO_2 occurs during winter, resulting in seasonal polar caps of dry ice that can be up to a few meters thick. The High Resolution Imaging Science Experiment (HiRISE) on-board MRO, observed the inactive Phoenix lander after a winter on the mid-northern latitudes of Mars and detected that one of its solar panels had been crushed due to the accumulation of a thick layer of CO_2 ice, providing in situ evidence for the latitudinal extent of the seasonal cap Diniega et al. (2021).

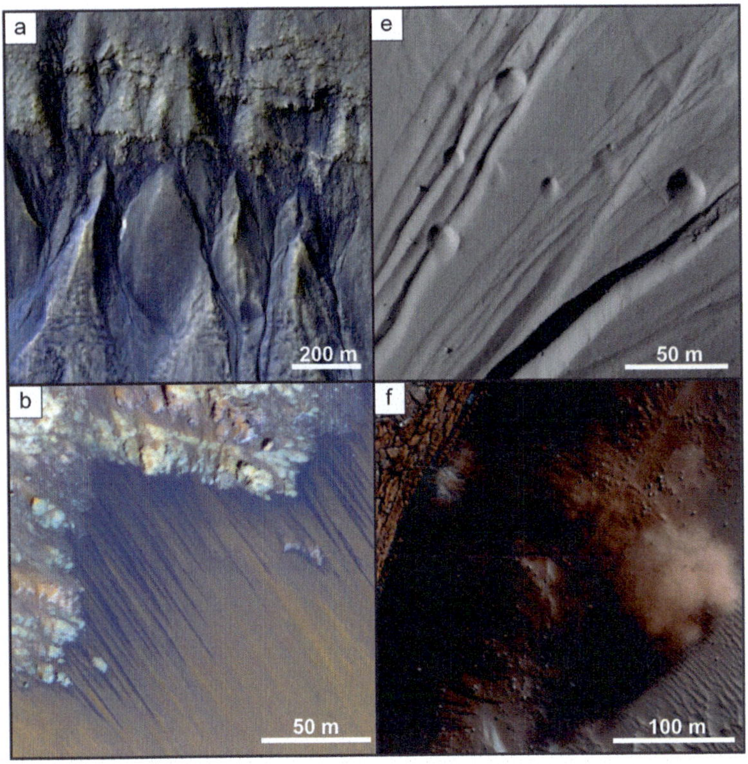

Fig. 7.1 HiRISE images of various ice and/or water-related landforms on Mars (with some discussed exceptions). **a** Gullies in a crater wall (ESP_014396_1415). **b** Recurring Slope Lineae in Coprates Chasma (ESP_034830_1670). **c** Araneiform terrain ("spiders") near the south polar caps (ESP_020914_0930). **d** "Swiss Cheese" terrain on the South Pole Residual Cap (ESP_012271_0940). **e** Pit craters and gullies in Hale Crater (PSP_003209_1445. **f** Avalanche in the North Polar Layered Deposits (ESP_016423_2640). **g** Blockfalls at the base of a polar cliff (ESP_036436_2645). **h** Dune alcoves in a north polar dune field (ESP_017768_2640). (https://www.uahirise.org/ NASA/JPL/University of Arizona)

7.1.2 Martian Epochs and Evolution

The geological history of Mars has been divided into three main Periods, each named after a region of Mars: Noachian (4.1–3.7 billion years ago), Hesperian (3.7–2.9 billion years ago), and Amazonian (2.9 billion years ago to present) (Scott and Carr 1978). Further geological mapping and stratigraphic studies also allowed subdividing each period into smaller Epochs, that is, Early, Middle, and Late Noachian; Early and Late Hesperian; and Early Middle and Late Amazonian (Tanaka 1986). An earlier Pre-Noachian period (4.5–4.1billion years ago) has also been identified (Frey 2005, 2006; Nimmo and Tanaka 2005).

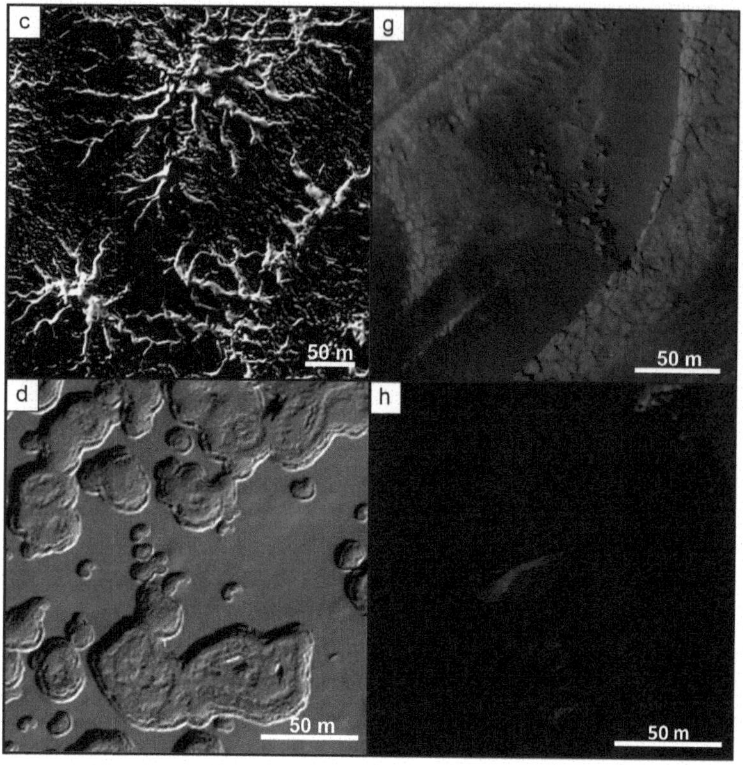

Fig. 7.1 (continued)

There is a consensus that early Mars (Noachian and Hesperian) had abundant water in all states that was responsible for many ancient features seen today (e.g., valley networks, outflow channels, the paleolake basins, and delta fans). Mineralogical alteration and past glaciation features further attest to surface modification by water. Estimates of the amount of water necessary to emplace all the sediments related to these features range from 100 s to 1000 m GEL (Global Equivalent Layer; Lasue et al. 2019 and references therein).

During the Amazonian, the environment at Mars became dry and cold, with very low surface atmospheric pressures and little to no liquid water. Today, there is no evidence of large volumes of liquid water near the surface, except for some disputed evidence of subsurface brine layers near the South Pole (e.g. Mattei et al. 2022; Lauro et al. 2021; Sori and Bramson 2019; Orosei et al. 2018). Instead, Mars has widespread H_2O ice, as ground ice, and as seasonal frost, and in its polar caps being the South Polar Cap the single location with the most water ice (Becerra et al. 2019). Hydrated minerals and water vapor are also present, but any liquid water that exists would most likely be transient due to deliquescence (Dundas et al. 2021).

This chapter focuses on analyzing the relationship between illumination conditions and volatile-related geomorphological processes, through examples of the applicability of the MarsLux code (Spagnuolo et al. 2019). The motivation is to highlight the potential of this tool in the field of planetary geomorphology, to showcase maps and examples of the capability of MarsLux, and to explore the relationship between illumination/energy and geomorphic features, especially those related to ice sublimation/condensation processes. After presenting the data and methods, we show two examples of impact craters (Louth crater on the Northern plains and Hale crater on the Southern Hemisphere) using MarsLux, and we place these studies in the context of a more global analysis.

7.2 Methods and Data

The MarsLux tool (Spagnuolo et al. 2019) is used to generate maps of illumination of different areas of Mars. This code is based on the geometric relationship between the position of the Sun and a point of interest (PoI) on the surface of Mars, taking into account the surrounding topography. The code allows the user to simulate the instantaneous illumination conditions. The user can start by choosing two dates and a time step, and the result can be an integer over that given time span. We present a modification to the original code, in which the orbital parameters are taken from Allison (1997) and Allison and McEwen (2000), following Mars24 equations. Instead, here we use a modified version that allows the user to tune the obliquity parameter. This version computes the horizontal coordinates of the Sun from an arbitrary point on Mars' surface. To this end, the position of Mars with respect to the Sun is first obtained using an approximate ephemeris that allows the computation of the Martian cartesian coordinates in the ICRF (Standish and Williams 1992). The routine then computes the angles that specify the orientation of the Mars figure also in the ICRF, i.e., the right ascension and declination of the planet's north pole, and its proper rotation angle, based on the fit given by Archinal et al. (2018). The first two angles are then transformed to the obliquity and the ascending node of the body equator of date with respect to the Earth's Equinox at J2000.0. This computation is as follows.

Let $\overrightarrow{R} = (R + h; \lambda; \varphi)$ be the position vector of a point on Mars' surface, where R is the reference radius of Mars, h is the height above R, λ is the longitude and φ is the latitude (these values are obtained from a relevant Digital Elevation Model). If $\overrightarrow{r_\odot}$ is the position vector of the Sun with respect to Mars, and $\overrightarrow{r}$ is the position vector of Mars with respect to the Sun, then it is evident that: $\overrightarrow{R} + \overrightarrow{r} + \overrightarrow{r_\odot} = \overrightarrow{0}$, which allows the calculation of the components of $\overrightarrow{r_\odot}$. Before performing the computation of the rectangular components of $\overrightarrow{r_\odot}$, the components of $\overrightarrow{r}$ must be transformed to the equatorial reference frame of Mars. Finally, the components of $\overrightarrow{r_\odot}$ are transformed to the local horizontal system by three consecutive rotations, from which the azimuth and altitude of the Sun for a specific point on Mars' surface are computed for a specific date and obliquity.

The simulation is divided into three steps or separate functions; the first step comprises the calculation of the Sun position in relation to the POI. This code outputs a five-column matrix for each POI, which, together with the DEM, is used for the evaluation of lighting conditions. With these instantaneous illumination maps, we estimate the different statistical parameters of the amount of light and energy. In a similar way to studies of solar panel arrays (e.g., Crisp et al. 2004), illumination was determined 24 times during each Sol, or once every Martian hour (~3700 s). The illumination maps generated represent an instantaneous solar zenith angle with respect to each topographic point of the input topography for the time step chosen.

With these instantaneous maps, the user can integrate over time and compute several statistical parameters to quantify the amount of light and energy received by each pixel, within the Region of Interest (RoI), in the established time period. The parameters obtained are total number of hours of light (full light if the POI is hit by the whole solar disk or partial if it is only hit by a fraction of the solar disk); daily accumulated amount of light intensity; daily maximum light intensity; and daily average of light intensity. All of these parameters can also be computed weekly and yearly if needed. The light intensity is obtained through Eq. 7.1:

$$S_i = S_o * \cos\cos(Z) * (D_m/D)^2 \tag{7.1}$$

where S_i represents the light intensity in W/m^2 corrected by the solar zenith angle Z and distance to the Sun (D). The instantaneous, spectrally integrated solar Fux at the top of the Martian atmosphere is $S_o = 590$ W/m^2 when the sun is at its mean distance of $D_m = 1.52$ AU. This result does not take into account any atmospheric correction and does not discriminate among different wavelengths. The daily average of energy is expressed in kilowatts per hour per area (kWh/m^2), which is traditionally used in solar power estimations (Acton 1996). To introduce a very basic model of atmospheric illumination correction, we recalculate the energy at the surface (O) by incorporating the atmospheric optical depth (τ) and the incidence angle (Z), as shown in Eq. 7.2:

$$O = S_i \exp(-\tau/\cos\cos(Z)) \tag{7.2}$$

We verify the results from the code by running a simulation for the Phoenix lander. We calculated the illumination received by the area of the solar panels along 160 Sols (starting on May 25, 2008). The results (Fig. 7.2) show that the code accurately calculates the known illumination conditions. The uncertainties observed are likely a consequence of factors like dust or frost deposits, clouds, or variations in opacity along the period analyzed.

The current solutions for Mars' orbital evolution (Fig. 7.3) are accurate between −20 Ma and +10 Myr and are routinely used for paleoclimate studies (Laskar et al. 2004). Orbital solutions also show that the mean obliquity between 5 and 20 Ma ago was ~35° Fluctuated, in the last 20 Myrs, between 15 and 45°. From these solutions, it is well established that at ~4.2 Ma, Mars entered a low obliquity epoch (Laskar et al.

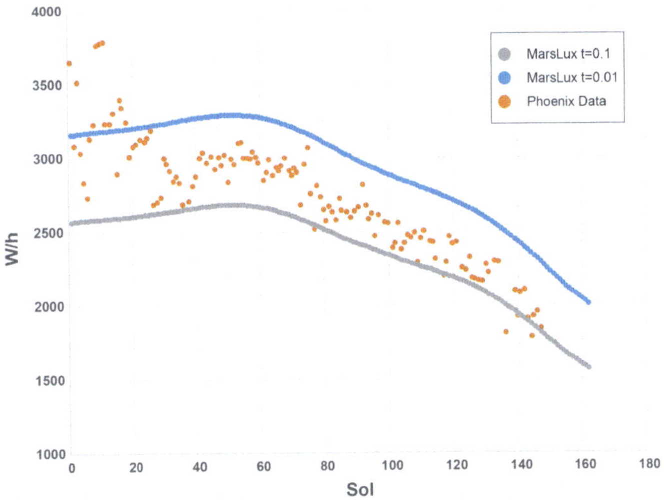

Fig. 7.2 Real energy data received by the Phoenix lander's solar panels (dots) (data taken from Coyne et al. 2009) compared to two results obtained by Marslux using different atmospheric opacities

2004; Smith et al. 2020). Orbital parameters, in particular obliquity, may generate many excursions away from the long-term average conditions.

In short computational times (a day to a month), depending on the resolution of the DEM used, MarsLux can create accurate illumination maps that can be translated into energy budget maps, for scales that range from local (meter scale) to global, and for any obliquity conditions and the other orbital parameters dependent on the specific date.

Using this approach only allows to study hypothetical situations and we are not modeling specific moments of Mars. Nevertheless, considering the chaotic nature of obliquity but the cyclical behavior of eccentricity, it is expected that some conditions

Fig. 7.3 Temporal evolution of the obliquity of Mars (data taken from Laskar et al. 2004). The black dot indicates the actual Martian obliquity. The red area represents the envelope of the high-frequency variations along the time

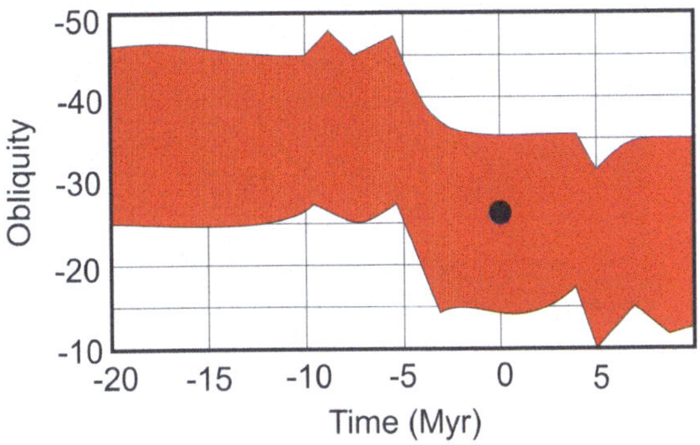

repeated over time. For this reason, we calculated insolation data for extreme obliquities, like 40° and 10° together with current obliquity conditions (25.19°) but the other orbital parameters remain as the current conditions. For example, we found that similar orbital values of current eccentricity and the argument of perihelion but with an obliquity of ~40° were found ~12 Myr ago based on the calculations of Laskar et al. (2004).

As in most recent climate/geomorphic studies, we will use the Mars calendar (Clancy et al. 2000; Piqueux et al. 2015a) to reference Martian time. This calendar uses the position of Mars along its orbit (given by LS, the areocentric longitude of the Sun from Mars) to measure time. Mars is at aphelion (distance from the sun ~1.664 AU) at Ls = 70°, near the northern Summer solstice, and at perihelion (distance from the sun ~1.384 AU) at Ls = 250°, near the southern summer solstice. Ls = 0° represents the northern vernal equinox and thus marks the beginning of spring in the northern hemisphere. Similarly, Ls = 90°, 180°, and 270° correspond to northern Summer solstice, Autumn equinox, and Winter solstice, respectively. Mars Year 1 (MY) therefore began arbitrarily on Ls = 0°, on April 11, 1955. For this chapter, we did all calculations for MY30 that started on October 26, 2009, and ended on April 8, 2011.

In the following sections, we present different study cases where we show illumination maps that could serve as base maps for geomorphic analysis. We first introduce a global analysis and then two local cases: Louth Crater in the Northern Hemisphere and Hale Crater in the Southern Hemisphere (Fig. 7.4). For the analysis, we used MOLA (Smith et al. 2001) and HRSC DEMs (Gwinner et al. 2016). As we calculate the instantaneous direct lighting conditions at every Martian hour, it is possible to therefore integrate our solutions to generate mean annual illumination maps, separate solutions by seasons, days, or any other arrangement of the data that is desired.

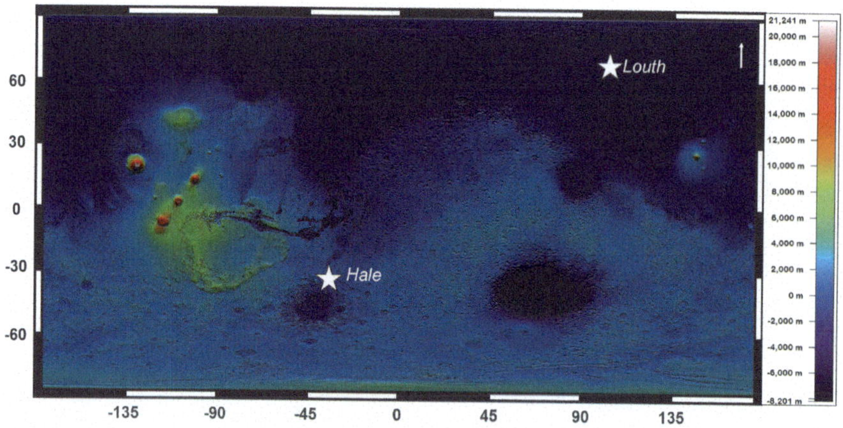

Fig. 7.4 Location of Louth and Hale craters analyzed in Sect. 7.3. (Topographic base map from https://pds-geosciences.wustl.edu/missions/mgs/mola.html)

7.3 Study Cases

7.3.1 Global Illumination

We first evaluate the global illumination conditions across the whole planet. In this section, we present the results for three different obliquities. For these maps, we used a global MOLA DEM (Smith et al. 2001) subsampled to 1 degree per pixel. Due to this subsampling several small features are lost in the analysis and the result is similar to a flat surface. Nevertheless, these global maps might be useful for comparison between the general illumination patterns, and the distribution of different geomorphic features. Figures 7.5 and 7.6 show the global illumination conditions as the total amount of energy received in MY30 modeled for three obliquities: 10° (Figs. 7.5a and 7.6a), 25.19° (Figs. 7.5b and 7.6b), and 40° (Figs. 7.5c and 7.6c). They also show the location of the landforms from Fig. 7.1 indicated.

As expected, the general pattern of illumination follows latitudinal bands, such that the more highly illuminated areas are near the Equator, and illumination diminishes towards the poles.

With an obliquity of 10° Mars' surface would have very low illumination conditions (Figs. 7.5a and 7.6a) resulting in less energy received over a year. The distribution of energy would be more symmetrical than today with an obliquity of 25.19°.

The current obliquity of 25.19° results in the Southern Hemisphere that receives less energy than the Northern Hemisphere (Figs. 7.5b and 7.6b).

In the higher obliquity scenario (40°), the planet has a more homogeneous illumination pattern, with significantly smaller energy differences between the poles and the equator (Figs. 7.5c and 7.6c). The global illumination map, when modeling the planet with an obliquity of 40°, shows that the equatorial band is on average "colder", which means that it receives less illumination, compared with obliquities of 10° and 25.19°.

In the present Martian scenario (Figs. 7.5b and 7.6b), with an obliquity of 25.19°, the equatorial region receives, on average, more illumination than the cases when the planet has both a higher and a lower obliquity. When comparing with some geomorphic features, it can be observed that active gullies are located mostly in the southern hemisphere where the energy would have the largest differences between a high obliquity scenario (Figs. 7.5c and 7.6c) and current conditions (Figs. 7.5b and 7.6b).

7.3.2 Louth Crater

In the Martian northern and southern circumpolar regions, there are over 40 ice deposits contained in the interior of craters as ice "icelands" (Sori et al. 2019; Conway et al. 2012). One of these deposits is found in Louth Crater. This crater is 36 km

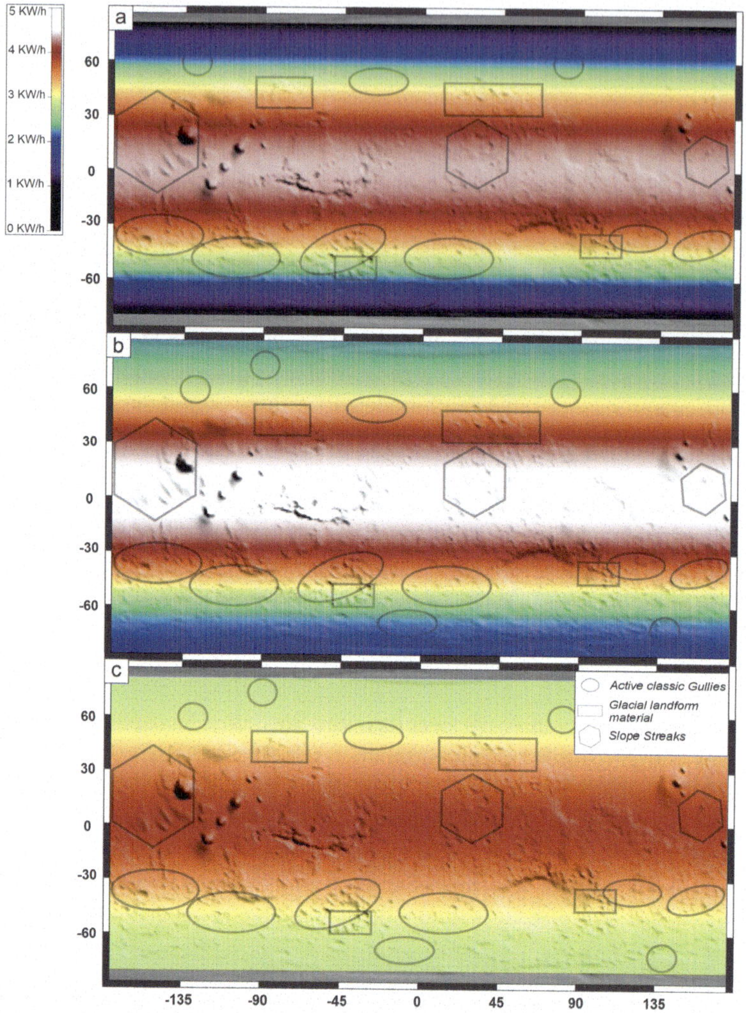

Fig. 7.5 Global maps of daily average energy compared to ice-related and active landforms (Location of landforms taken from: Diniega et al. 2021; Dundas et al. 2021, and Levy et al. 2014). **a** 10° obliquity; **b** 25.19° obliquity; **c** 40°, obliquity. Polygons show the location of active classic gullies, glacial material, and slope streaks. (Topographic base map from https://pds-geosciences.wustl.edu/missions/mgs/mola.html)

wide and 1.5 km deep. Located at ~70.2°N/102.3°W, it is one of the southernmost water ice "island" in the North hemisphere. Its interior contains a ~10 km diameter ice mound emplaced near the center of the crater floor, slightly shifted toward the northern wall. Previous work has described the morphology and composition of the ice mound (Brown et al. 2008), as well as modeled its mass balance (Bapst et al. 2018). However, the origin of the ice deposit is still poorly understood. Prior work proposed

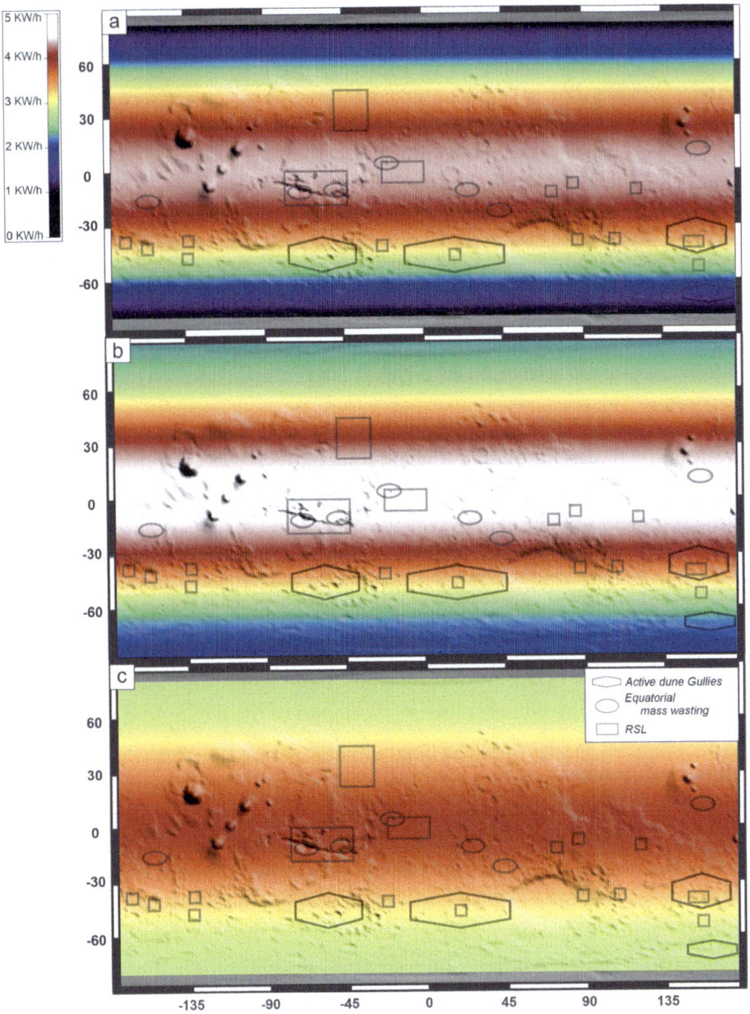

Fig. 7.6 Continuation of Fig. 7.5: Polygons show the location of active gullies, equatorial mass wasting processes, and Recurrent Linear Slopes (RSL) (Topographic base map from https://pds-geosciences.wustl.edu/missions/mgs/mola.html)

that inner-crater ice mounds developed under unique atmospheric conditions in what are known as "cold traps" (Conway et al. 2012). An alternative hypothesis argues that these equatorward ice deposits might be related with a more extensive polar ice cap that then retreated (Fishbaugh and Head 2000). However, most geological evidence points to a model of atmospheric deposition of water vapor in the center of the crater (Brown et al. 2008). Moreover, the offset position of the ice mound was attributed to these depositional processes, as well as aeolian activity and sublimation (Brown et al. 2008).

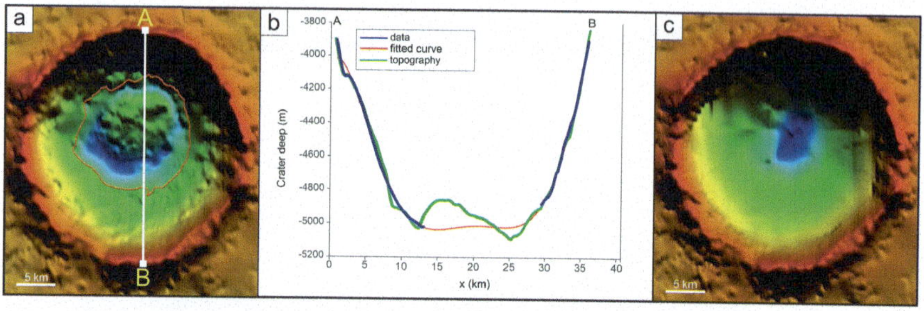

Fig. 7.7 **a** Louth crater DEM. We generated several profiles across different azimuths (e.g. N–S) and then "cut" the DEM using a contour encircling the ice mound. **b** Example of a topographic profile. **c** Reconstructed topography of an "empty" Louth crater

We explore the relationship that illumination and solar energy might have with the position of the ice mound inside the Louth crater using the MarsLux tool. We look at different scenarios of high and low obliquity in an empty crater, and present a series of maps that could help discern between three hypotheses of the ice mound formation: (i) it represents residual North Polar Cap deposits; (ii) the ice mound was formed through direct deposition in the center of Louth crater and shifted toward the north; and (iii) it was formed through direct deposition at its present offset position from the center. We analyzed the distribution of shadows and solar illumination within Louth Crater in order to evaluate whether the northern part of the ice mound represents a low illumination region that enhances ice precipitation by a self-shadowing feedback. For the illumination analysis, we used the MOLA DEM (Smith et al. 2001) downsampled to 600 m/px.

To analyze the effects of topography we generate a new DEM without the ice mound (with the same resolution) but with a typical bowl-shaped crater floor ("Empty Crater"). To create this new topography, we first mapped the limits of the ice mound using HiRISE and CTX images. Then, we extracted these points from the original topographic grid. Finally, using an interpolating polynomial technique for different topographic profiles, we reconstruct the crater floor, filling the grid again with the interpolated points (Fig. 7.7).

Using the topography and given orbital parameters, the code calculates shadows and illumination for each pixel for a period of time. First, we calculated the illumination in the present conditions, using the actual topography and a Martian obliquity of 25.19°, and then we analyzed the reconstructed empty crater.

7.3.2.1 Present-Day Louth Crater (with Mound)

During one Martian year (16516 h), some parts of the interior of Louth crater were in darkness for a maximum of ~9500 h. Figure 7.8 shows that the shadows are as expected for a northern crater, but some areas stand out for the anomalously high amount of time that they spend in shadow. At present time, the area enclosed between

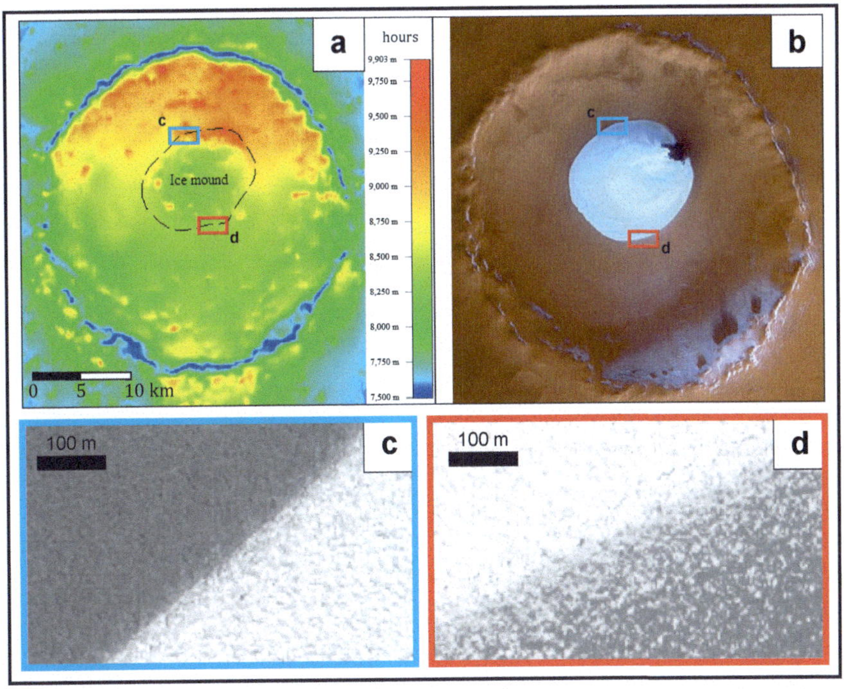

Fig. 7.8 **a** Analysis of total hours of darkness during MY30. **b** HRSC image of Louth crater (North is up; HRSC 234192, ESA/DLR/FU Berlin (G. Neukum), CC BY-SA 3.0 IGO). **c** and **d** northern and southern edges of the ice mound, respectively (ESP_027017_2505, https://www.uahirise.org/ NASA/JPL/University of Arizona)

the mound and the northern crater wall has the lowest average energy and spends the most amount of time in full darkness. We can thus infer that there is a clear "self-shadowing" process of this area of the crater by the ice mound (Fig. 7.8). This self-shadowing effect may be influencing the morphological differences in the ice-mound edges. The northern boundary of the ice mound is abrupt, while the southern boundary is diffuse, with patchy, light-toned material, likely water ice, extending out hundreds of meters from the ice mound (Bapst et al. 2018).

7.3.2.2 Reconstructed Crater (No Mound)

For the reconstructed crater, we analyzed the illumination at different obliquities (10°, 25.19°, 40°, and 60°) (Fig. 7.9). The results of the illumination map do not show any large anomalies in the interior, except for the northern interior edge of the rim, which receives less energy than the southern region of the crater interior in the high obliquity cases (Fig. 7.9b–d). This pattern is reversed in the low obliquity scenario (10°) where the southern side of the rim receives less energy than the northern. As

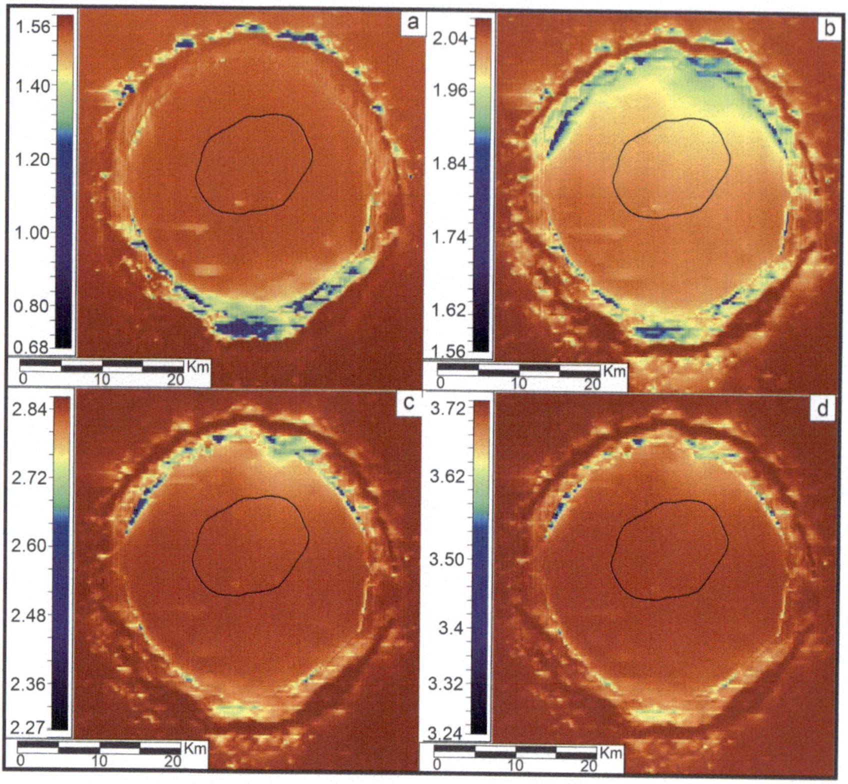

Fig. 7.9 Mean daily energy received by Louth crater as calculated by Eqs. 7.1 and 7.2 in kW/h for different obliquities (using $\tau = 0.01$): **a** 10°; **b** 25.19 (present-day) **c** 40° and **d** 60°. The black contour represents the current position of the ice mound

seen in Fig. 7.9a, lower obliquities result in lower amounts of energy received on average in the whole area.

Figure 7.9b shows the illumination of an empty Louth crater in the current obliquity (29.19) Two highlights can be pointed out compared with the results obtained for other obliquities. First, the interior received on average less energy than the surroundings. When modeling with other obliquities, the interior of the crater receives almost the same amount of energy as the outside of the crater. Second, the interior shows the largest disparity in energy between the northern and southern areas and the location of the present-day ice mound (black outline in the figure) seems to correlate with this disparity. This correlation is not observed with other obliquities (Fig. 7.9a, c, and d).

7.3.3 *Hale Crater*

Hale crater is a large complex crater formed in the source region of Uzboi Vallis, along the northern rim of the Argyre basin, within the Nereidum Montes region. It is one of the largest Martian impact craters known to have generated channels that radiate outward from the impact site (El-Maarry et al. 2013). The interior of Hale crater contains lots of evidence of ice-related landforms, such as rock glaciers, moraines, solifluction lobes, thermokarsts, and polygonal patterned terrain (Hauber et al. 2011; Soare et al. 2019), and gullies related to ice melting processes and CO_2 seasonal frosts (Reiss et al. 2009).

We discuss the presence of ice/water-related landforms in Hale Crater based on the illumination pattern within the crater. Using CTX base images and HiRISE, we identified several ice-related landforms such as rock glaciers and moraines. In this example, we also analyzed one Martian year (MY30) making a calculation every Martian hour. For this example, we used a different approach. After generating the maps of average daily energy, maximum darkness, and cycles between light and darkness, we implemented a k-means algorithm in order to identify areas with similar characteristics (Fig. 7.10).

We also incorporated in the analysis some topographic layers like slope maps and altitude. This implementation was done with a cluster analysis in R (Maechler et al. 2019) and the result is a final map with five classes based on the six input maps. These five classes correspond to specific illumination parameters and topographic characteristics. The results from the illumination analysis compared with the map of the location of ice-related landforms show a correlation between them (Fig. 7.11), implying therefore a correlation between solar fluxes, topographic characteristics, and specific Landforms. Areas with poor illumination conditions (blue and green classes in Fig. 7.10a) are dominated by ice-related features like rock glaciers and gullies (Fig. 7.10b, c). In areas with maximum illumination (yellow and red classes), no distinguishable ice-related morphologies are observed. In addition, the crater has three distinctive dune fields (Jones et al. 2011) that seem to be located in areas of changing conditions, i.e., in the limits between the red and orange classes. These areas represent limits between regions with more and less average solar flux, respectively. Finally, the pitted material that dominates the crater interior corresponds mainly to the red (maximum illumination) class.

7.4 Final Comments

In this chapter, we discuss the relationship between illumination and geomorphic features (and thereby the processes that formed them) by means of a global map and two local examples of illumination conditions derived from orbital parameters and topography.

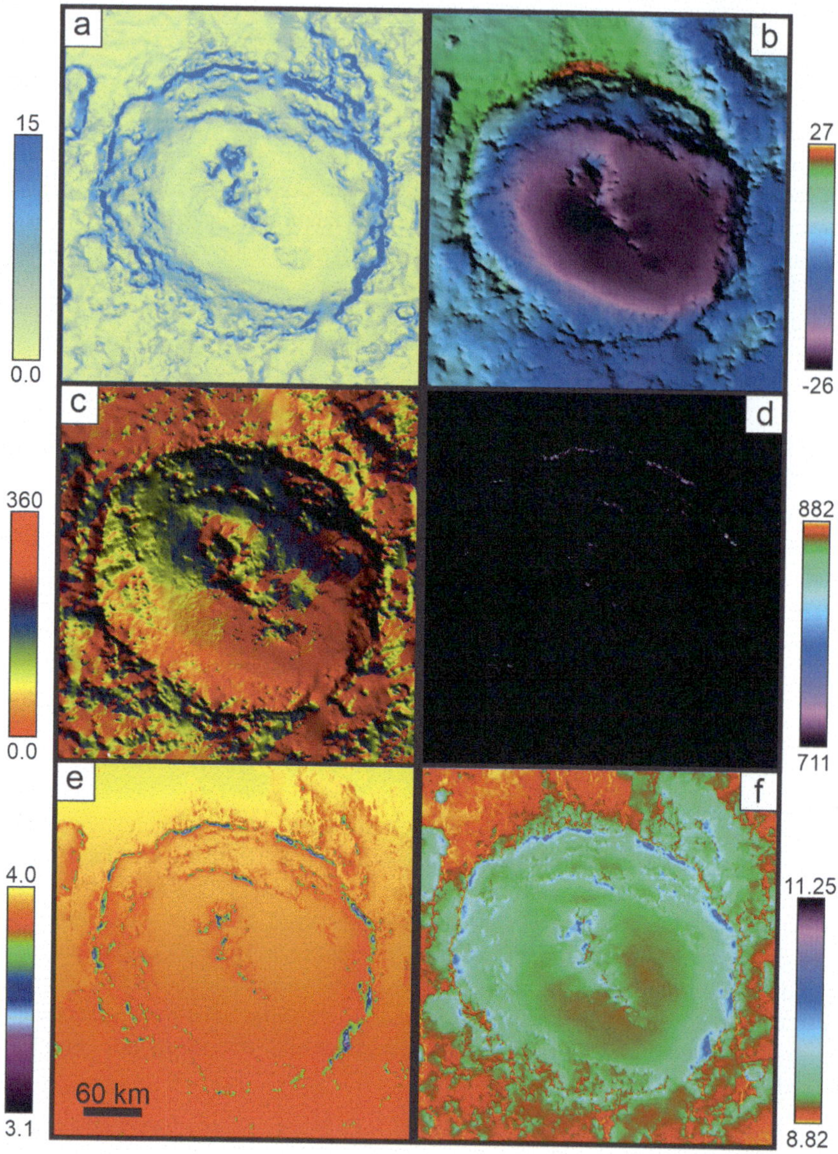

Fig. 7.10 Layers used in the k-mean analysis: **a** Slope (degrees), **b** altitude (km), **c** aspect (degrees), **d** cycles of light and darkness, **e** average daily energy (kWh), **f** maximum hours of darkness (hours)

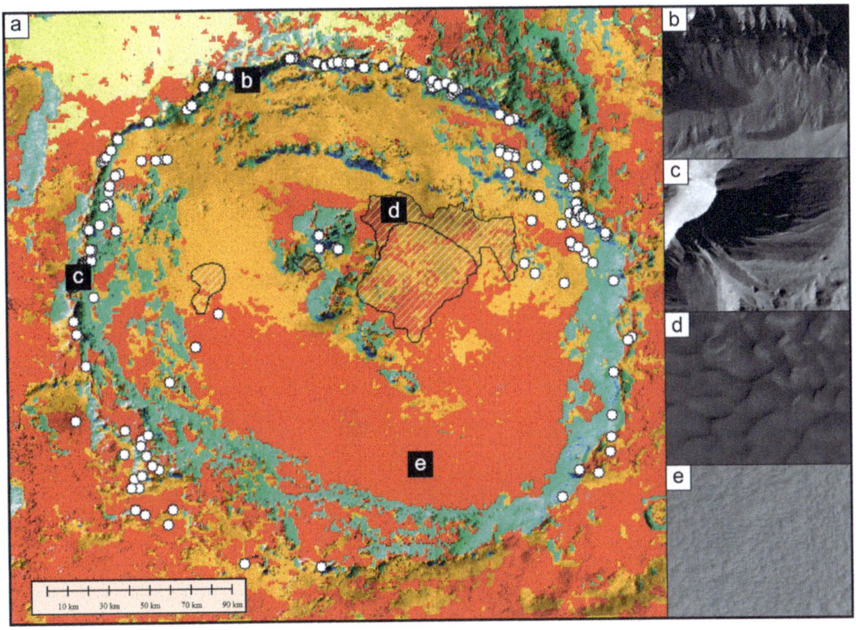

Fig. 7.11 **a** Map of classes identified by the k-means algorithm using the layers from Fig. 7.10 (see text for colour-coded). White dots represent identified ice-related morphologies and hatched terrain dune fields. **b** Rock glaciers deposit with polygonal scarps, **c** Gullies, **d** Dunes, **e** Pitted terrain. (https://www.uahirise.org/NASA/JPL/University of Arizona)

The Martian global pattern of illumination, calculated for different obliquities sheds interesting results and leads to several questions. Could the current geographic distribution of active geomorphic features be related to the symmetrical illumination received? Are mass-wasting processes in the equatorial regions influenced by the past presence of ice in these areas (during the 40° mean obliquity epoch)? Head et al. (2003) proposed that "with obliquity >45° (6 Myr ago), the polar sublimation rate is very high, and ice is predicted to persist at the equator." These changes in global illumination may also affect the global wind pattern and circulation of the atmosphere. With the same atmospheric pressure values, it is possible to speculate that winds are stronger today than those 4 My ago when the global illumination was more homogeneous.

The analysis of Louth crater also gives rise to questions about the relationship between the illumination conditions in the past and the origin of the ice cap inside the crater. The differences in solar flux between the 40° mean obliquity epoch and present-day are contradicted by the colder interior of the crater in the past. One could speculate that if the ice cap formed by condensation directly in its current location inside the crater, then it could only have started forming after the shift to a lower mean obliquity around 5 My ago (Fig. 7.3).

Hale crater contains a plethora of glacial and periglacial deposits. We showed that there is a correlation between these features and illumination "classes" (i.e., particular patterns of illumination and topographic conditions). Are the current conditions in equilibrium with these glacial deposits or not?

Establishing a direct link between geomorphic processes and illumination conditions, and thereby providing an answer to the above questions is beyond the scope of this chapter, and requires further and in-depth analysis and modeling. Our intention here is only to present the illumination maps as a tool for further research devoted to understanding Martian geomorphology in the past and present.

Acknowledgements We thank anonymous reviewers for constructive and insightful comments, and the editors for efficient editorial handling. Thanks to the funds of PIP 11220200101675CO (CONICET) and PICT 3697-01796 (FONCyT) for which this work was possible.

References

Acton CH (1996) Ancillary data services of NASA's Navigation and Ancillary Information Facility. Planet Space Sci 44(1):65–70. https://doi.org/10.1016/0032-0633(95)00107-7

Allison M (1997) Accurate analytic representations of solar time and seasons on Mars with applications to the Pathfinder/Surveyor missions. Geophys Res Lett 24(16):1967–1970. https://doi.org/10.1029/97GL01950

Allison M, McEwen M (2000) A post-Pathfinder evaluation of areocentric solar coordinates with improved timing recipes for Mars seasonal/diurnal climate studies. Planet Space Sci 48(2–3):215–235. https://doi.org/10.1016/S0032-0633(99)00092-6

Archinal BA, Acton CH, A'Hearn MF, Conrad A, Consolmagno GJ, Duxbury T, Hestroffer D, Hilton JL, Kirk RL, Klioner SA, McCarthy D, Meech K, Oberst J, Ping J, Seidelmann PK, Tholen DJ, Thomas PC, Williams IP (2018) Report of the IAU working group on cartographic coordinates and rotational elements: 2015. Celest Mech Dyn Astr 130(3):22. https://doi.org/10.1007/s10569-017-9805-5

Bapst J, Byrne S, Brown AJ (2018) On the icy edge at Louth and Korolev craters. Icarus 308:15–26. https://doi.org/10.1016/j.icarus.2017.10.004

Becerra P, Sori MM, Thomas N, Pommerol A, Simioni E, Sutton SS et al (2019) Timescales of the climate record in the south polar ice cap of Mars. Geophys Res Lett 46:7268–7277. https://doi.org/10.1029/2019GL083588

Brown A, Byrne S, Tornabene L, Roush T (2008) Louth crater: evolution of a layered water ice mound. Icarus 196(2):433–445. https://doi.org/10.1016/j.icarus.2007.11.023

Clancy RT, Sandor BJ, Wolff MJ, Christensen PR, Smith MD, Pearl JC, Conrath BJ, Wilson RJ (2000) An intercomparison of ground-based millimeter, MGS TES, and Viking atmospheric temperature measurements: seasonal and interannual variability of temperatures and dust loading in the global Mars atmosphere. J Geophys Res 105(E4):9553–9571. https://doi.org/10.1029/1999JE001089

Conway SJ, Hovius N, Barnie T, Besserer J, Le Mouélic S, Orosei R, Read NA (2012) Climate-driven deposition of water ice and the formation of mounds in craters in Mars' north polar region. Icarus 220(1):174–193. https://doi.org/10.1016/j.icarus.2012.04.021

Coyne J, Jackson W, Lewicki C (2009) Phoenix electrical power subsystem—Power at the Martian Pole. In: 7th International energy conversion engineering conference. American Institute of Aeronautics and Astronautics, Denver, Colorado

Crisp D, Pathare A, Ewell RC (2004) The performance of gallium arsenide/germanium solar cells at the Martian surface. Acta Astronaut 54(2):83–101. https://doi.org/10.1016/S0094-5765(02)00287-4

Diniega S, Smith IB (2020) High-priority science questions identified at the Mars workshop on Amazonian and present-day climate. Planet Space Sci 182:104813. https://doi.org/10.1016/j.pss.2019.104813

Diniega S, Bramson AM, Buratti B, Buhler P, Burr DM, Chojnacki M, Conway SJ, Dundas CM, Hansen CJ, McEwen AS, Lapôtre MGA, Levy J, Mc Keown L, Piqueux S, Portyankina G, Swann C, Titus TN, Widmer JM (2021) Modern Mars' geomorphological activity, driven by wind, frost, and gravity. Geomorphology 380:107627. https://doi.org/10.1016/j.geomorph.2021.107627

Dundas CM, Becerra P, Byrne S, Chojnacki M, Daubar IJ, Diniega S, Hansen CJ, Herkenhoff KE, Landis ME, McEwen AS, Portyankina G, Valantinas A (2021) Active Mars: a dynamic world. J Geophys Res: Planets 126(8):e2021JE006876. https://doi.org/10.1029/2021JE006876

El-Maarry MR, Dohm JM, Michael G, Thomas N, Maruyama S (2013) Morphology and evolution of the ejecta of Hale crater in Argyre basin, Mars: results from high resolution mapping. Icarus 226(1):905–922. https://doi.org/10.1016/j.icarus.2013.07.014

Fishbaugh KE, Head JW (2000) North polar region of Mars: topography of circumpolar deposits from Mars Orbiter Laser Altimeter (MOLA) data and evidence for asymmetric retreat of the polar cap. J Geophys Res 105(E9):22455–22486. https://doi.org/10.1029/1999JE001230

Frey HV (2005) Impact constraints on the age and origin of the lowlands on Mars. Geophys Res Lett 33:L08S02. https://doi.org/10.1029/2005GL024484

Frey HV (2006). Impact constraints on, and a chronology for, major events in early Mars history. J Geophys Res 111:E08S91. https://doi.org/10.1029/2005JE002449

Gwinner K, Jaumann R, Hauber E, Hoffmann H, Heipke C, Oberst J, Neukum G, Ansan V, Bostelmann J, Dumke A, Elgner S, Erkeling G, Fueten F, Hiesinger H, Hoekzema NM, Kersten E, Loizeau D, Matz KD, McGuire PC, Mertens V, Michael G, Pasewaldt A, Pinet P, Preusker F, Reiss D, Roatsch T, Schmidt R, Scholten F, Spiegel M, Stesky R, Tirsch D, van Gasselt S, Walter S, Wählisch M, Willner K (2016) The high resolution stereo camera (HRSC) of mars express and its approach to science analysis and mapping for mars and its satellites. Planet Space Sci 126:93–138

Hauber E, Reiss D, Ulrich M, Preusker F, Trauthan F, Zanetti M, Hiesinger H, Jaumann R, Johansson L, Johnsson A, Van Gasselt S, Olvmo M (2011) Landscape evolution in Martian mid-latitude regions: insights from analogous periglacial landforms in Svalbard. Geol Soc London Spec Publ 356(1):111–131. https://doi.org/10.1144/SP356.7

Head JW, Mustard JF, Kreslavsky MA, Milliken RE, Marchant DR (2003) Recent ice ages on Mars. Nature 426(6968):797–802. https://doi.org/10.1038/nature02114

Jones AP, McEwen AS, Tornabene LL, Baker VR, Melosh HJ, Berman DC (2011) A geomorphic analysis of Hale crater, Mars: the effects of impact into ice-rich crust. Icarus 211(1):259–272. https://doi.org/10.1016/j.icarus.2010.10.014

Kieffer HH (2013) Thermal model for analysis of Mars infrared mapping: KRC THERMAL MODEL. J Geophys Res Planets 118(3):451–470. https://doi.org/10.1029/2012JE004164

Knapmeyer M, Stähler SC, Daubar I, Forget F, Spiga A, Pierron T, van Driel M, Banfield D, Hauber E, Grott M, Müller N, Perrin C, Jacob A, Lucas A, Knapmeyer-Endrun B, Newman C, Panning MP, Weber RC, Calef FJ, Böse M, Ceylan S, Charalambous C, Clinton J, Dahmen N, Giardini D, Horleston A, Kawamura T, Khan A, Mainsant G, Plasman M, Lemmon M, Lorenz R, Pike WT, Scholz J-R, Lognonné P, Banerdt B (2021) Seasonal seismic activity on Mars. Earth Planet Sci Lett 576:117171. https://doi.org/10.1016/j.epsl.2021.117171

Laskar J, Correia ACM, Gastineau M, Joutel F, Levrard B, Robutel P (2004) Long term evolution and chaotic diffusion of the insolation quantities of Mars. Icarus 170(2):343–364. https://doi.org/10.1016/j.icarus.2004.04.005

Lasue J, Clifford SM, Conway SJ, Mangold N, Butcher FEG (2019) The hydrology of Mars including a potential cryosphere. In: Volatiles in the Martian Crust. Elsevier, pp 185–246

Lauro SE, Pettinelli E, Caprarelli G, Guallini L, Rossi AP, Mattei E, Cosciotti B, Cicchetti A, Soldovieri F, Cartacci M, Di Paolo F, Noschese R, Orosei R (2021) Multiple subglacial water bodies below the south pole of Mars unveiled by new MARSIS data. Nat Astron 5(1):63–70. https://doi.org/10.1038/s41550-020-1200-6

Leighton RB, Murray BC (1966) Behavior of carbon dioxide and other volatiles on Mars: a thermal model of the Martian surface suggests that Mars's polar caps are solid carbon dioxide. Science 153(3732):136–144. https://doi.org/10.1126/science.153.3732.136

Levy JS, Fassett CI, Head JW, Schwartz C, Watters JL (2014) Sequestered glacial ice contribution to the global Martian water budget: Geometric constraints on the volume of remnant, midlatitude debris-covered glaciers: buried martian glaciers. J Geophys Res Planets 119(10):2188–2196. https://doi.org/10.1002/2014JE004685

Maechler M, Rousseeuw P, Struyf A, Hubert M, Hornik K (2019) Cluster analysis basics and extensions. R package version 2.1.0

Martínez GM, Newman CN, De Vicente-Retortillo A, Fischer E, Renno NO, Richardson MI, Fairén AG, Genzer M, Guzewich SD, Haberle RM, Harri A-M, Kemppinen O, Lemmon MT, Smith MD, de la Torre-Juárez M, Vasavada AR (2017) The modern near-surface Martian climate: a review of in-situ meteorological data from Viking to Curiosity. Space Sci Rev 212(1–2):295–338. https://doi.org/10.1007/s11214-017-0360-x

Mattei E, Pettinelli E, Lauro SE, Stillman DE, Cosciotti B, Marinangeli L, Tangari AC, Soldovieri F, Orosei R, Caprarelli G (2022) Assessing the role of clay and salts on the origin of MARSIS basal bright reflections. Earth Planet Sci Lett 579:117370. https://doi.org/10.1016/j.epsl.2022.117370

Nimmo F, Tanaka K (2005) Early crustal evolution of Mars. Ann Rev Earth Planet Sci 33:133–161

Orosei R, Lauro SE, Pettinelli E, Cicchetti A, Coradini M, Cosciotti B, Di Paolo F, Flamini E, Mattei E, Pajola M, Soldovieri F, Cartacci M, Cassenti F, Frigeri A, Giuppi S, Martufi R, Masdea A, Mitri G, Nenna C, Noschese R, Restano M, Seu R (2018) Radar evidence of subglacial liquid water on Mars. Science 361(6401):490–493. https://doi.org/10.1126/science.aar7268

Piqueux S, Byrne S, Kieffer HH, Titus TN, Hansen CJ (2015a) Enumeration of Mars years and seasons since the beginning of telescopic exploration. Icarus 251:332–338. https://doi.org/10.1016/j.icarus.2014.12.014

Piqueux S, Kleinböhl A, Hayne PO, Kass DM, Schofield JT, McCleese DJ (2015b) Variability of the martian seasonal CO_2 cap extent over eight Mars years. Icarus 251:164–180. https://doi.org/10.1016/j.icarus.2014.10.045

Piqueux S, Kleinböhl A, Hayne PO, Heavens NG, Kass DM, McCleese DJ, Schofield JT, Shirley JH (2016) Discovery of a widespread low-latitude diurnal CO_2 frost cycle on Mars. J Geophys Res Planets 121:1174–1189. https://doi.org/10.1002/2016JE005034

Reiss D, Hiesinger H, Hauber E, Gwinner K (2009) Regional differences in gully occurrence on Mars: a comparison between the Hale and Bond craters. Planet Space Sci 57(8–9):958–974. https://doi.org/10.1016/j.pss.2008.09.008

Scott DH, Carr MH (1978) Geologic map of Mars. US Geological Survey Miscellaneous Investigations Series Map I-1083, scale 1:25,000,000

Smith DE, Zuber MT, Frey HV, Garvin JB, Head JW, Muhleman DO, Pettengill GH, Phillips RJ, Solomon SC, Zwally HJ, Banerdt WB, Duxbury TC, Golombek MP, Lemoine FG, Neumann GA, Rowlands DD, Aharonson O, Ford PG, Ivanov AB, Johnson CL, McGovern PJ, Abshire JB, Afzal RS, Sun X (2001) Mars Orbiter Laser Altimeter: experiment summary after the first year of global mapping of Mars. J Geophys Res 106(E10):23689–23722. https://doi.org/10.1029/2000JE001364

Smith MD (2008) Spacecraft observations of the Martian atmosphere. Annu Rev Earth Planet Sci 36(1):191–219. https://doi.org/10.1146/annurev.earth.36.031207.124334

Smith IB, Hayne PO, Byrne S, Becerra P, Kahre M, Calvin W, Hvidberg C, Milkovich S, Buhler P, Landis M et al (2020) The Holy Grail: a road map for unlocking the climate record stored within Mars' polar layered deposits. Planet Space Sci 184:104841

Soare RJ, Conway SJ, Gallagher C, Dohm JM, Reiss D (2019) Periglacial complexes and the deductive evidence of 'wet'-flows at the Hale impact crater, Mars. Geol Soc London Special Publ 467(1):211–231. https://doi.org/10.1144/SP467.7

Sori MM, Bramson AM (2019) Water on Mars, with a grain of salt: local heat anomalies are required for basal melting of ice at the South Pole today. Geophys Res Lett 46(3):1222–1231. https://doi.org/10.1029/2018GL080985

Sori MM, Bapst J, Becerra P, Byrne S (2019) Islands of ice on Mars and Pluto. J Geophys Res: Planets 124:2522–2542. https://doi.org/10.1029/2018JE005861

Spagnuolo MG, Carballo FD, Marco Figuera R, Rossi AP (2019) MarsLux: HI-resolution illumination maps generator for Mars. Earth Space Sci 6(1):146–155. https://doi.org/10.1029/2018EA000403

Spiga A, Forget F (2009) A new model to simulate the Martian mesoscale and microscale atmospheric circulation: validation and first results. J Geophys Res 114:E02009. https://doi.org/10.1029/2008JE003242

Standish EM, Williams EM (1992) ORBITAL ephemerides of the sun, moon, and planets. In: P.K., S (ed) Explanatory supplement to the astronomical Almanac, University Books, Mill Valley, CA, pp 279–374

Tanaka KL (1986). The stratigraphy of Mars. Proceedings of the Lunar and planetary science conference 17, part 1. J Geophys Res 91:E139–E158

Author Index

A. Coronato and G. B. Alves (eds.), *Latin American Geomorphology*, The Latin American Studies Book Series, https://doi.org/10.1007/978-3-031-55178-9

GPSR Compliance

The European Union's (EU) General Product Safety Regulation (GPSR) is a set of rules that requires consumer products to be safe and our obligations to ensure this.

If you have any concerns about our products, you can contact us on ProductSafety@springernature.com

In case Publisher is established outside the EU, the EU authorized representative is:

Springer Nature Customer Service Center GmbH
Europaplatz 3
69115 Heidelberg, Germany

The manufacturer's authorised representative in the EU is Springer Nature Customer Service Centre GmbH, Europaplatz 3, 69115 Heidelberg, Germany. If you have any concerns regarding our products, please contact ProductSafety@springernature.com

Printed and bound by CPI Group (UK) Ltd, Croydon, CR0 4YY
23/09/2026
02221541-0007